Contents

List of Figures

List of Tables

ix

List of Contributors

Hans-Georg Bohle
Professor of Geography, University of Heidelberg, Südasien Institut, Abteilung Geographie, Im Neuenheimer Feld 330, D-69120 Heidelberg, Germany.

Florian Dünckmann
Geography (Ph. D.), University of Kiel, Geographisches Institut, Ludewig-Meyn-Strasse 14, D-24098 Kiel, Germany.

Helmut Geist
Geography (Ph. D.), German-American Center for Visiting Scholars, 1607 New Hampshire Avenue, NW, Washington, D.C. 20009, United States of America.

Barbara Göbel
Anthropology (Ph. D.), University of Bonn, Seminar für Völkerkunde, Römerstrasse 164, D-53117 Bonn, Germany.

Thomas Krings
Professor of Development Studies, University of Freiburg, Institut für Kulturgeographie, Werderring 4, D-79085 Freiburg i. Br., Germany.

Fred Krüger
Professor of Geography, University of Erlangen-Nürnberg, Geographie, Kochstr. 4, D-91054 Erlangen, Germany.

Beate Lohnert
Geography (Ph. D.), University of Osnabrück, Fachgebiet Geographie, Seminarstrasse 19, D-49069 Osnabrück, Germany.

Matthias K. B. Lüdeke
Physics (Ph. D.), Potsdam Institute for Climate Impact Research, Integrated Systems Analysis Department, P.O. Box 601203, D-14412 Potsdam, Germany.

Gerhard Petschel-Held
Physics (Ph. D.), Potsdam Institute for Climate Impact Research, Integrated Systems Analysis Department, P.O. Box 601203, D-14412 Potsdam, Germany.

Fritz Reusswig
Sociology (Ph. D.), Potsdam Institute for Climate Impact Research, Global Change and Social Systems Department, P.O. Box 601203, D-14412 Potsdam, Germany.

Sven Schade
Geography (Ph. D.), University of Bayreuth, Geographisches Institut, Universitätsstrasse 30, D-95440 Bayreuth, Germany.

Veronika Ulbert
Geography (Ph. D.), Friedrich-Ebert-Stiftung, Stabsabteilung, Godesberger Allee 149, D-53175 Bonn, Germany.

Foreword

One of the major objectives of DFG's research priority programme on the social dimensions of global environmental change (SPP 194)[1] was - and still is - the provision of an organizational framework for researchers from various fields of study to join task forces. Among the five task forces organized for the total of three periods covering 1995 to 2000, most of the contributions in this volume directly originate from the activities of Task Force 5 entitled Environmental Perception and Coping Strategies in Endangered Ecosystems of the Developing World. A minor part of the contributions prove to be strongly related to the SPP activities of task force 5 in that non-DFG affiliated researchers from related fields either got irrepressibly attracted by and heavily involved into task force activities (by simply having attended one or all of the DFG's numerous colloquia and workshops) or partly even worked under close supervision of SPP members under their own budget.

While task force 5 started as a group of social researchers coming from a wide field of scientific disciplines (including geography, anthropology, sociology and demography), from the very first set up of the organizational framework it was clear that - different from other task forces - the special focus will not only be a thematical one, but more so explicitly related to empirical investigations among social groups in different areas in what is commonly known as the developing world. More so, from numerous meetings and discussions since about 1996 it emerged that special attention is well worth to be given to the notions of environmental perception, (un)sustainability, livelihood systems and vulnerability by relating them to the more action oriented and strategic aspects of global environmental change - with particular reference to what natural scientists use to call endangered ecosystems - however, in next to all cases not giving a proper format to the human dimensions involved.

At the highlighting meeting in Osnabrück in May 1998, there was common ground originating from the observation that one of the task force's major objectives since 1995, i.e., categorizing research results as originating from individual SPP projects, should be left behind. It was agreed upon that in future a further and rather daring step should be tried towards aggregating the combined efforts aimed at the integration of views upon ecosystem, perception and social strategy in the form of a publication as presented here. The emphasis so far given to what could be called the meso and micro level of social analysis, i.e., nations, regions, markets, state, social groups and even individual households, was not seen to be a major obstacle when dealing with the world-wide (or macro) dimensions of changing environments, but more so a necessary drive towards scaling down both the processes and effects of global (environmental) change. The point has to be made here that the special emphasis in this volume upon - and the particular devotion given in individual chapters to - the social or human dimensions of environmental change is seen to easily shape the bulk of future activities in a field of research that has longly been dominated by the natural scientists' perspective (and access to funds). Together with the compelling need to strengthen the regional component, and adhering to Thomas Kuhn's (1962) notion with abandon, the editors would never have run into this editorial enterprise if not convinced that 'paradigmatic change happens'.[2]

It should be mentioned that, if not stated otherwise, all translations from sources originally written in German and cited in the text fall under the full responsibility of the editors as it is with any omissions or typographical errors.

Beate Lohnert, Mazabuka and Osnabrück
Helmut Geist, Neuss and Washington, DC

Notes

[1] Access to further information on SPP 194 (Schwerpunktprogramm Globale Umwelt-veränderungen: Sozial- und verhaltenswissenschaftliche Dimensionen) of the Deutsche Forschungsgemeinschaft (DFG) is provided by internet (in English). The homepage including links, products, task forces, objectives, participants, projects, events and news is <http://www.psychologie.uni-freiburg.de/umwelt-spp-eng/welcome.html>, while further information on task force 5 can be found via <http://www.uni-freiburg.de/ umwelt-spp-eng/spp-tafo/task5.html>.

[2] This has lately been expressed by one of the speakers at the Fifth Scientific Advisory Council Meeting of the International Geosphere-Biosphere Programme (A Study of Global Change) at UNEP's headquarter in Nairoibi. As a matter of fact, both the shift from natural to human dimensions as well as towards strengthening the regional emphasis could be read from the lips of numerous contributors attending SAC V; see Geist, H. (1998), 'Africa and Global Environmental Change: The Shift from Natural to Human Dimensions, A report on the 5th Scientific Advisory Council Meeting (SAC-V), Nairobi, September 1-7, 1998', afrika spectrum, vol. 33 (2). Thus, attendants willing to do so could realize how Thomas Kuhn's notion of paradigmatic change materialized; Kuhn, T. (1962), The Structure of Scientific Revolutions, University of Oxford Press, Oxford.

Acknowledgements

Editors and contributors know all too well their indebtedness to various collaborators and supporters. Since they were brought together by activities or related contacts within the first German Research Priority Programme (*Schwerpunktprogramm*, SPP 194) on 'Social Dimensions of Global Environmental Change' (*Mensch und globale Umweltveränderungen*), much and next to all of the indebtedness is owed to the *Deutsche Forschungsgemeinschaft* (DFG) in Bonn. Being the central public funding organization for academic research in Germany (thus comparable to a Research Council or Foundation in other countries) and paying special attention to the education and support of young scientists and scholars, in 1995 the DFG set SPP 194 in motion for a two year period and with most of the individual projects aligned even gambled a second and third phase later.

We are extremely grateful to the programme's academic director Dr. Manfred Niessen and the steering as well as scientific advisory committees' members for enabling us to proceed with our work. And we salute the patience and professionalism of the people at Ashgate Publishing Limited who suffered ever-changing titles, missing artwork, and numerous excuses with unflappable good will.

At the respective departments of national universities where individual research projects were attached to, special thanks go to many individuals from secretarial offices and cartographic laboratories. In particular, Torsten M. Schlautmann, University of Osnabrück, has earned more thanks than his salary could reflect by keeping sharp eyes to copy-editing uncountable drafts of all eleven chapters and by now being an old hand at producing books.

A sterling staff has made all the difference, but so has the opportunity to work with contributors whose intellectual commitment held sway over more lucrative opportunities. That commitment brought many of them to Osnabrück in the spring of 1998 to launch the book project. For that gathering, we are indebted to the DFG for its continued sponsorship as well as to the Department of Geography in Osnabrück for the local arrangements and unmatched hospitality.

As contributors and researchers we gratefully acknowledge the benevolent support of the national scientific research councils in various countries of the developing world (Nepal, Laos, South Africa, Tanzania, Malawi, Botswana, Brazil, Argentina, Dominican Republic) for granting research clearance and thus contributing to the enhancement of knowledge on how to cope with environmental change and the endangerment of livelihood systems. Included here are the numerous institutions further related to research clearance and involved in the progress of field surveys to be carried out.

The contributors owe dinner to colleagues, partners, research assistants, friends and family members not only for understanding and support when being away from their home place, but also for support and assistance during their stay in the respective countries and areas under study. In the order of appearance, and as laid down in the single case studies of the book, special thanks are expressed as follows (while the views represented in the eleven chapters of the volume are those of the authors and not of any institution listed above or below).

Hans-Georg Bohle, University of Heidelberg, to the DFG.

Thomas Krings, University of Freiburg, to the DFG and former director of the Lao-German Forestry team, Professor Brechtel, for assistance, advice and accommodation during his stay in Laos in spring 1995.

Beate Lohnert, University of Osnabrück, to the DFG and many individuals based at the University of Cape Town in South Africa. Special thanks go to Prof. Andrew ('Mugsy') Spiegel from the Department of Social Anthropology for asking the right questions at the right time, to Prof. Sue Parnell, Sophie Oldfield and Darryll Kilian from the Department of Environmental and Geographical Sciences for fruitful discussions, and to Prof. Heinz Rüther, Dr. Scott Mason and Mike Barry from the Department of Geomatics for their assistance in getting hold of aerial photographs.

Helmut Geist, German-American Center for Visiting Scholars, to the DFG and his African partners, Frank F.Y. Matumula in Malawi and A. M. Mlingi in Tanzania, for monitoring the field surveys. Further acknowledged are grants by the World Bank's Human Development Network for a short-term consultancy under commitment no. A35186 ('Tobacco Control Policies in the Developing World').

Sven Schade, University of Bayreuth, to Deutsche Gesellschaft für Technische Zusammenarbeit (GTZ) for project funding under title no. 89.2143.9-01.142 of the 'Tropical Ecology Support Programme'. Further acknowledged is the support of the Land Management Programme of the Swedish International Development Agency (SIDA) and Handeni

Integrated Agroforestry Project (GTZ) when carrying out field survey in Tanzania.

Fred Krüger, University of Erlangen-Nürnberg, to the DFG.

Florian Dünckmann, University of Kiel, to the DFG and to Instituto Florestal (IF), Secretaria do Meio Ambiente do Estado de Sao Paulo (SMA), Universidade de Sao Paulo (USP) and Instituto de Terras (IT) for providing support when carrying out the field survey.

Barbara Göbel, University of Bonn, to the DFG, Auswärtiges Amt, Deutscher Akademischer Austauschdienst and Fondation Fyssen. Special thanks go to the people of Huancar for their friendship and co-operation during fieldwork.

Veronika Ulbert, formerly University of Freiburg and now Friedrich Ebert Stiftung (FES), to Peter Asmussen and Dr. Martin Schneichel from Proyecto Bosque Seco (Dominican Republic, GTZ) for assistance during the field survey. Further thanks go to Wolf G. Schnellbach, Tracy Asmussen and Joachim Scheiner for their help in getting the paper formated in time.

Gerhard Petschel-Held, *Matthias Lüdeke* and *Fritz Reusswig*, Potsdam Institute for Climate Impact Research (PIK), would like to thank Oliver Moldenhauer and Martin Cassel-Gintz for their support in modelling and presenting the modelling output as well as Stephen Sitch for comments on an earlier version of the manuscript.

Being more than a gesture of respect, this book is devoted to people met in the areas under study and having to cope with environmental change under terms of mitigation so widely different from the ones the contributors find in their home country to be at hand. Be it in countries of Asia, Africa or Latin America, partners of dialogue being full of understanding were found who spent precious time of patiently responding to questions despite of their day-to-day struggles in getting along for themselves. For that reason, honest hopes are expressed in that the findings presented here will not be put aside into book shelves of university libraries but more so find their way to a broader audience of the environmental education, policy and planning community in the so-called developing as well as developed worlds, thus contributing to bridge the divide since there is only one world to live in.

In this context, and to start with materializing our aspirations for distribution beyond the scope of *academia*, we acknowledge the interest raised among and responded by the Potsdam Institute for Climate Impact Research (PIK) being at the heart of providing scientific input to the German Government's Advisory Council on Global Change (WBGU). In particular, Fritz Reusswig from PIK's Department of Global Change and

Social Systems has truly earned more than one dinner for his agreement to take over the challenging job of drawing conclusions from the case studies presented. Special thanks go to Matthias Lüdeke and Gerhard Petschel-Held for contributing to the concluding chapter which thus came to adopt a fully transdisciplinary and methodology-oriented view.

List of Abbreviations

ADB	Asian Development Bank
ADDO	Arusha Diocese Development Organization
BMU	Bundesministerium für Umwelt, Naturschutz und Reaktorsicherheit
CBS	Central Bureau of Statistics
CCM	Chama cha mapinduzi (Party of National Unity)
CIFOR	Center for International Forestry Research
CIPROS	Centro de Investigación y Promoción Social
CONATEF	Comisión Nacional Técnica Forestal
CPR	Common Property Resources
DARG	Developing Areas Research Group
DBSA	Development Bank of Southern Africa
DFG	Deutsche Forschungsgemeinschaft
EAMP	Environmental Assessment and Management Plan
ECOSOC	Economic and Social Council of the United Nations
EGAT	Electricity Generating Authority of Thailand
ESA	Earth System Analysis
FAO	Food and Agriculture Organization of the United Nations
FEBROSUR	Federación de Productores y Productoras del Bosque Seco
FES	Friedrich Ebert Stiftung
FFS	Food Self-Sufficiency
FORESTA	National (Dominican) Forest Administration
GAIM	Global Analysis, Interpretation and Modelling
GCTE	Global Change and Terrestrial Ecosystems
G(E)C	Global (Environmental) Change
GIS	Geographical Information System
GTOS	Global Terrestrial Observing System
GTZ	Deutsche Gesellschaft für Technische Zusammenarbeit
IAD	Instituto Agrario Dominicano
ICIDI	Independent Commission on International Development Issues
ICOLD	International Commission on Large Dams
ICSU	International Council of Scientific Unions
IDGC	Institutional Dimensions of Global Environmental Change Project

IDRC	International Development Research Center
IDS	Institute of Development Studies
IF	Instituto Florestal
IGBP	International Geosphere-Biosphere Programme
IGU	International Geographical Union
(I)HDP	(International) Human Dimensions Programme on Global Environmental Change
IIED	Institute for Environment and Development
IMF	International Monetary Fund
IPCC	Intergovernmental Panel on Climate Change
IT	Instituto de Terras
ITC	International Institute for Aerial Survey and Earth Sciences
ITGA	International Tobacco Growers' Association
KAS	Konrad Adenauer Stiftung
IUCN	International Union for Conservation of Nature
LAMP	Land Management Project of Simanjiro District
LAPC	Land and Agriculture Policy Centre
LLDC	Least Developed Countries
LUCC	Land-Use and Land-Cover Change
LWADD	Liwonde Agricultural Development Division
MAB	Man and the Biosphere Programme
MFDP	Ministry of Finance and Development Planning
MRC	Mekong River Commission
MSDF	Metropolitan Spatial Development Framework
NBCA	National Biodiversity Conservation Area
NDP	National Development Plan
NEM	New Economic Mechanisms
NEMC	National Environment Management Council
NFS	National Food Strategy
NGO	Non-Governmental Organization
NGS	National Geographic Society
NKGCF	German National Committee on Global Change Research
NPC	National Planning Commission
NRDP	National Rural Development Programme
NT(EC)	Nam Theun (Electricity Consortium)
PAB	Programma de Acción de Bosque Seco
PDR	People's Democratic Republic (of Laos)
PIK	Potsdam Institute for Climate Impact Research
PLA	Participatory Learning on Action
PR	Private Resources

PRA	Participatory Rural Appraisal
QDE	(Concept of) Qualitative Differential Equations
RDP	Rural Development Programme
RIDEP	Ruvuma Integrated Development Scheme
RRA	Rapid Rural Appraisal
SAC	Scientific Advisory Council
SADC	Southern African Development Community
SALDRU	Southern Africa Labour and Development Research Unit
SAMCU	Songea Agricultural Marketing Cooperative Union
SARDC	Southern African Research and Documentation Centre
SEADE	Fundacão Sistema Estadual de Análise de Dados
SIDA	Swedish International Development Agency
SMA	Secretaria do Meio Ambiente do Estado de Sao Paulo
SODA	Songea Development Action
SPP	Schwerpunktprogramm (Research Priority Programme)
SRF	Swaminathan Research Foundation
TCP	Tarangire Conservation Project
TRIM	Tobacco Research Institute of Malawi
UN	United Nations
UNCED	United Nations Conference on Environment and Development
UNCHS	United Nations Centre for Human Settlements (HABITAT)
UNCTAD	United Nations Conference on Trade and Development
UNDP	United Nations Development Programme
UNEP	United Nations Environment Programme
UNSECO	United Nations Educational, Scientific and Cultur Organization
UNRISD	United Nations Research Institute for Social Development
UNU	United Nations University
URT	United Republic of Tanzania
USP	Universidade de São Paulo
WBGU	German Scientific Advisory Council on Global Change
WCED	World Commission on Environment and Development
WESGRO	Association for the Promotion of the Western Cape's Economic Growth
WHO	World Health Organization of the United Nations

1 Endangered Ecosystems and Coping Strategies
Towards a conceptualization of environmental change in the developing world

BEATE LOHNERT AND HELMUT GEIST

Introduction

Against the background that 'coping with global environmental change has come to appear one of humankind's most pressing problems' (Meyer and Turner, 1995), the chapter attains to explore how this relates to situations which face the majority of people living at present in the neighbourhoods and villages of urban and rural Africa, Asia and Latin America. By doing so, we are gambling with the notion that a certain hesitance could be observed to undertake this type of research (Ehlers and Krafft, 1998).

The chapter further attains to introduce into some of the outcomes of the first German research priority programme on the social dimensions of global environmental change (GEC). The nine regional case studies presented in the volume originate from DFG's (Deutsche Forschungsgemeinschaft) task force 5 that has adopted a particular view on resource use and livelihood systems in endangered ecosystems of the developing world (see foreword for more programme and project related information).

In structural terms, social sciences have drawn heavily on the interpretation of case studies to seek some insights into the main factors modifying human societies and natural environments, and so do the contributions collected in the volume. The distribution of case studies among the continents as delineated in Figure 1.1 happened to be a result of individual research interests rather than falling under the structural intentions of the programme (or the editors), though probably reflecting some steering activities from the side of the scientific advisory committee of the DFG. From striking similarities of the social, economic and political aspects involved in the cases from Nepal, Laos, South Africa, Tanzania, Malawi, Botswana, Brazil, Argentina and the Dominican Republic, we recognize in

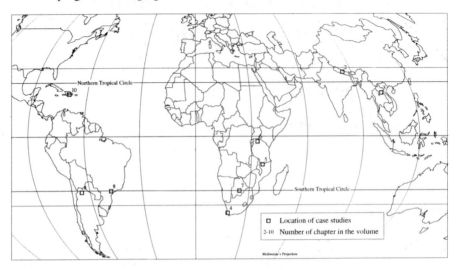

Figure 1.1   Location of case studies in the volume

the framework of political ecology a suitable platform which is well worth exploring further and get linked to social GEC research.

By looking upon political ecology the way we do, it is not attained at all to rank various other approaches lower, e.g., ecosystem or landscape analysis (Heal, Menaut and Steffen, 1993; Walker and Steffen, 1996), eco-restructuring (Manshard, 1998 a), industrial, urban or social metabolism or ecology (Fischer-Kowalski, Haberl, Hüttler, Payer, Schandl, Winiwarter and Zangerl-Weisz, 1997), or Earth System Analysis (ESA) (Schellnhuber and Wenzel, 1998). However, we recognize in political ecology a conceptual framework that captures characteristic aspects of human-environmental relations specifically applicable to the study of developing world situations.

Following the nine individual cases, a view upon them from the concept of syndromes and the ESA perspective of transdisciplinary and integrated systems analysis as adopted by the Potsdam Institute for Climate Impact Research (PIK) is provided in the concluding chapter.

Environmental Change and Endangered Ecosystems

According to the Scientific Advisory Committee (SAC) of the International Geosphere-Biosphere Programme (A Study of Global Change) and IGBP's think tank Global Analysis, Interpretation and Modelling (GAIM), from now on 'in less than a decade the scientific community could begin to realise fully coupled, dynamical (prognostic) models of the Earth System' (Moore, 1996). This relates to the 'coupling across four critical interfaces, i.e., terrestrial ecosystems and the atmosphere, the ocean and the atmosphere, the chemistry of the atmosphere and the physics of the atmosphere, and terrestrial ecosystems and the coastal oceans'. As it had earlier been with human responses to climate change, i.e., when social science advice was asked for only to find out how people (should) respond, at the recent SAC V meeting in Nairobi in 1998, it was asked what will be, given our understanding of the couplings among physical and biochemical aspects of the earth system, the form of its future interactions with human activities.

From handling social climate impact research as such one could recognize that most of social GEC research tends to continue to focus on how environmental change should be managed, rather than on why or by whom. By further reviewing the present and foreseeable conceptualizations of the social dimensions of GEC, it is obvious that they are not sensitive at all towards Third-world situations. Though it was stated, for example, by Noble (1996) that IGBP work 'is a challenging task in itself ... (but) it will only be a partial success unless we are able to take into account human-related driving forces', in an attempt to link human dimensions to landscape dynamics, he drew information from the responses of less developed agricultural societies to climate change in the past calling the subchapter 'What we can learn from the past'. The ambiguities of such a procedure are instantaneously obvious, and the author implicitly recognized this in terms of 'significant changes' that have occurred such as 'cash cropping, international trade, famine relief (and) externally subsidized warfare' (Noble, 1996). These arguments, however, are deeply rooted in the political-economy sphere of analysing environmental change and point out the significance to connect the natural sphere to political economy what perfectly lies at the heart of political ecology.

What Type of Endangerment?

With regard to threatened environments, it was clearly pointed out by Kasperson, Kasperson, Turner, Dow and Meyer (1995) that so far any 'de-termination of what constitutes a state of criticality has been largely subjective and judgemental, with no searching exploration of the associated conceptual and methodological issues'. However, be it academic or popular writing, the endangerment of ecosystems remains a central topic and continues to pervade much of the social dimension literature on environmental change (as it is in this volume, too).

At this point we suggest to turn to a recent attempt of summarizing the 'properties of ecosystems that have implications for institutions' (say, social dimensions). Reference is made to the International Human Dimensions Programme (IHDP) component called Institutional Dimensions of Global Environmental Change Project (IDGC) stating that the 'approach seeks to endogenize the role of social institutions in large-scale biophysical systems, by looking at human systems as subsystems of the ecosphere' (Pritchard, Colding, Berkes, Svedin and Folke, 1998). According to IDGC, 'ecological disturbance play(s) an especially important role', and five socially important ecosystem properties are worked out to bridge the 'misfit between ecosystems and institutions'. These properties are (i) ecosystems as life support systems, (ii) ecosystems as 'dynamically variable systems', (iii) ecosystems as 'spatially heterogenous systems', (iv) ecosystems as 'complex, evolving systems', and (v) 'disturbance as an important part of development'.

To introduce the view of political ecology upon the suggested 'misfit between ecosystems and institutions', the notion of environmental degradation will briefly be raised here. It is taken from the seminal work of Blaikie and Brookfield (1987, p. 6) using the term 'capability'. In doing so, they link the human-environmental 'misfit' (or, in the words of political ecology, the constantly shifting dialectic between society and land-based resources) to a purely societal 'misfit' (i.e., the constantly shifting dialectic within classes and groups within society itself) what is the political ecology concern to conceptualize the issue.

> (D)egradation is defined as a reduction in the capability of land to satisfy a particular use. If land is transferred from one system of production or use to another, say, from hunter-gathering to agriculture, or from agricultural to urban use, a different set of its intrinsic qualities become relevant and provide the physical basis for capability. Land may be more or less capable in the new context. This is important, because it must not be supposed that deforestation necessarily constitutes degradation in a social sense, even though it certainly leads to changes in micro-climate, hydrology and soil. Socially, degradation must relate to capability (...).

Different from IDGC's conceptualization, none of the case studies presented in the volume will use the term or notion 'human subsystems of the ecosphere', but more so ground ecology in the web of social relations that ties households together, and links them to larger economic and political entities, namely, the market, access to assets, land tenure, systems of surplus extraction and the state. Though system analysis (or related thinking) holds strong power over researchers' minds, it is stressed here, by using the words of Watts (1985), that the approach as adopted by the case studies in the volume 'demands a careful study of local-level processes and demonstrates that environmental change needs to be carried out not in vague terms but at specific locations and among specific segments of ... society'.

Handling Geocentric and Anthropocentric Views

With the case studies adopting a social science point of view, it is not suggested to treat humans as external to ecosystems while the latter are given the function of a 'black box'. As could be seen from the environmental setting of each case – be it the semiarid drylands of the Maasai steppe (chapter 6) or of rural Botswana (chapter 7), be it the humid Atlantic rain forest of Southeast Brazil (chapter 8) or the Caribbean dry forest ecosystem (chapter 10) – the linkages, for instance, to functional biodiversity losses and related policies such as that of UNEP (1992), UNCED (1992) or BMU (1992) are made more or less explicit (chapters 4, 6 and 9). The same holds true for key structuring processes of GEC such as land cover transformation through large-scale deforestation (chapters 3, 8 and 10) and soil fertility decline (chapters 5 and 6).

However, next to any pertinent and highly ranked anthropological or ontological reasoning about 'humans being a factor of disturbance in the geosystem' (Hambloch, 1986), about 'human disturbance of world ecosystems' (Hannah, Lohse, Hutchinson, Carr and Lankerani, 1994), or about 'ecological disturbance playing an especially important role' (Pritchard, Colding, Berkes, Svedin and Folke, 1998) is avoided here. This is done in order to relate the social analysis of environmental change not so much to the properties of ecosystems but more so to the three aspects of sustainability that in the best of all cases could be balanced by the land managers, i.e., social, economic and ecological aspects (see in particular chapter 2). Most of the underlying concepts of endangerment will therefore have a strong regional component and more or less resemble the qualitative and holistic notion that endangerment involves overall human uses of the environment that appear to be unsustainable over the near term (this and the next generation) in the absence of radical adjustments in their scale and composition (Kasperson, 1992, 1993; Turner, Kasperson, Kasperson, Dow and Meyer, 1995).

Thus, geocentric views will hardly be recognizable in the case studies presented. As it was pointed out by Kasperson, Kasperson, Turner, Dow and Meyer (1995), the use of a purely natural science definition – i.e., human-induced perturbation altering the biophysical system to a substantially different, less diverse and more or less degraded one – 'would require the designation of most urban-industrial complexes and most major farming regions as critical or endangered environmental areas'. Instead, most of the indications of endangerment will be grounded here in the web of social relations so that flips or bifurcations in ecosystems are related, for instance, to changing modes of natural resource management such as from grazing to farming (chapter 6), to the (un)intentional invasion of genetically engineered crops into biodiverse agroecosystems such as tobacco in Southeast African highlands (chapter 5), or to the clash of indigenous and modern knowledge systems upon biodiversity matters such as in the case of a marginal, pastoral community in the Andean highlands of Argentina (chapter 9).

With regard to key concepts and issues of ecological endangerment (or criticality), an array of often overlapping or conflicting terms has arisen such as sensitivity, stability, susceptibility, adaptability, resistance, resilience, equilibrium, marginality, fragility and vulnerability (Kasperson, Kasperson, Turner, Dow and Meyer, 1995; Manshard, 1998 a). While some of the chapters will rely upon terms as specified by natural science literature, a recurring theme in the volume will be vulnerability in the polit-ecological context of livelihood security and environmental endangerment. Some of the chapters will mildly paraphrase (and thus acknowledge) the outcomes of natural science literature on endangerment of ecosystems, while none of the case studies will fall into the other extreme of adopting a rigid social science point of view. As could be seen from cases so different such as the mountain ecosystem of Nepal (chapter 2), the dryland ecosystem of rural Botswana (chapter 7) or the urban ecology of metropolitan Cape Town settled in one of the world's uniquest biodiversity zones (chapter 4), all cases are far from adopting a purely anthropocentric perspective in that environmental constraints on human activity are minimized and assumed away, or social and technological change is seen as potentially sufficient to deal with possible endangerments.

Though lacking a rigorous definition and means of measurement, the concept of ecosystems as life-support system, however, strikes at an essential quality of human-environment interactions and deserves full quotation. Odum (1989, p. 13) has defined the 'life-support environment' as 'that part of the earth that provides physiological necessities of life, namely, food and other energy mineral nutrients, air and water', while a 'life-support system' is seen to be the functional term for the environment, organisms, processes, and resources interacting to provide these physical necessities (...) In terms of the landscape, agricultural systems (plus) natural systems (constitute) life-support systems.

Kasperson, Kasperson, Turner, Dow and Meyer (1995) demonstrated how the concept has been adapted and expanded by social scientists under such umbrella terms as 'basic human needs' or 'human development indices', how it pervaded much of the rhetoric at the 1997 UN Earth Summit in Rio de Janeiro (Quarry, 1992), and how it was central to the Brundtland Commission's (WCED) arguments on 'our common future' (World Commission on Environment and Development, 1987).[1]

If one agrees upon the view of Kasperson, Kasperson, Turner, Dow and Meyer (1995) that among the varied threats to the global environment 'the most central (and most relevant to notions of criticality) is endangerment of the ability to sustain human life over the long term', the point has to be made here that many of the following case studies taken from the social context of the developing world will prove that the global and long-term concern materializes as an already pressing concern how to realize life chances at all and how to put modes of mere physical survival into practice. What about twenty years ago had been a distinct 'north-south' component of 'survival' in a commission report under the auspices of Willy Brandt (ICIDI, 1980) seems now to have turned to a blurred 'common future' rhetoric. And even IDGC's perception of what makes an ecosystem a 'life-support system' does not allow for a proper allocation of ecosystem properties to the specific conditions of deprivation and survival as prevalent in the 'south'.

While it is not attained at all to undervalue the institutional aspects outlined by IDGC, from an action-oriented perspective for survival and life chances in the developing world, however, any systemic configuration in terms of 'nature' and 'human subsystem' is avoided here, and a special focus given instead to the external and internal aspects of vulnerability of individuals or groups and the specific position of people or social groups in the total context of social relations.

To conclude with, and to emphasize the general view adopted and the individual research designs used in the case studies of the volume, i.e., mostly addresssing the household level, Watts (1985) has continued to argue that through the 'careful study (of) environmental change ... at specific locations and among specific segments of ... society (...) the social relations of production and exchange are central not only to understanding the complexities of land-use decisions but also to broaching the paradox of why – and for whom – the problem of environmental change arises at all'.

What Type of Environmental Change?

The debate about the intrinsic qualities of environmental change can be broken down by the pessimists' and optimists' point of view. While many observers believe that – though much remains unkown for a decade, or so, about the physical, chemical and biological processes involved – an environmental tragedy is looming and the earth system will be unable to function as a life-support system in the near future, others claim that, against the background of past experiences, the doomsday scenario is overplayed and human ingenuity will cope again. However, wide recognition could be assumed on the fact that 'earth (is) transformed by human action' (Turner, Clark, Kates, Richards, Mathews and Meyer, 1990) and certain to be transformed in future.

Given 'some general validity to the pessimists' arguments ('it is better to be a pessimist and be proved wrong than to be an optimist and be wrong'), their cause could be supported by observation of current trends within society which are increasing the nature and pace with which we are ravaging the earth. Different from the more optimistic views at the beginning of the twentieth century, many arguments of coping with GEC have to be 'remapped' at the end of the twentieth century as it is summarized by Johnston, Taylor and Watts (1995 a, p. 298).

> First, pressure on earth's resources is growing because of increasing population. Secondly, demands on those resources are increasing more rapidly than is the population because of greater per capita material expectations (...). Thirdly, very few parts of the earth's surface remain relatively untouched by these demands – the colonial 'escape valves' of previous centuries have been removed – and conversion of increasing tracts to non-productive uses, through urbanization processes, reduces the amount of land available for exploitation. Finally, technology advances have been such in recent decades that not only has our ability to ravage the earth been magnified manyfold but in addition we are ravaging it in new ways, through technologies that enter the core biological and chemical processes of life.

However, the patterns as well as sources and impacts of environmental change are not uniform across the globe, and so is a wide range of problems having a largely political structure.

Global environmental change   In an attempt to conceptualize trajectories of the earth transformed, it was noted by Meyer and Turner (1995) that environmental 'changes will find different expression and have different consequences in different regions, but they will be significant almost everywhere'. If the notion of GEC is to remain a useful distinction, then some intermediate definition is required since, on the one hand (and in the strictest sense), 'no change is fully global inasmuch as none occurs uniformly across the earth', while on the other hand (and in the most expansive sense), 'all change is global inasmuch as all changes are ultimately connected with one another through physical and social processes alike' (Meyer and Turner, 1995).

Table 1.1   Two types of global environmental change (GEC)

Type	Characteristic	Examples
Systemic	Direct impact on globally functioning system	Industrial and land use emissions of 'greenhouse' gases Industrial and consumer emissions of ozone-depleting gases Land-cover changes in albedo
Cumulative	Impact through worldwide distribution of change	Groundwater pollution and depletion Species depletion and genetic alteration (biodiversity)
	Impact through magnitude of change (share of global resource)	Deforestation Industrial toxical pollutants Soil depletion on prime agricultural lands

Source: Turner, B.L.II, Kasperson, R.E., Meyer, W.B., Dow, K.M., Golding, D., Kasperson, J.E., Mitchell, R.C. and Ratick, S.J. (1990), 'Two Types of Global Environmental Change: Definitional and Spatial-Scale Issues in their Human Dimensions', *Global Environmental Change*, vol. 1 (1), p. 15.

In Table 1.1, two types of the global impacts of human activity are provided, while the first ('globally systemic change') refers 'to the spatial scale of functioning of a system', and the second ('globally cumulative change') 'occurs on a worldwide scale, or represents a significant fraction of the total environmental phenomenon of global resource' (Turner, Clark, Kates, Richards, Mathews and Meyer, 1990).

Current academic as well as popular writing and interest in global change has so far arisen from the systemic impacts of human activitiy, namely, greenhouse climate change and ozone depletion. Most of the case studies collected in the volume, however, will focus on the cumulative aspects of environmental change. In particular, chapter 3 (on the multi-faceted process of forest transformation and depletion in Laos), chapter 4 (on the transformation of urban ecology milieux through massive squatting and in-migration in South Africa), chapter 5 (on the ecological impact of farming tobacco in African highlands), chapter 6 (on woodland depletion and conversion of grazing land into farmland in Tanzania), chapter 8 (on the creeping encroachment of protected humid wilderness areas at the Brazilian Atlantic coast) and chapter 10 (on the scope of dry forest transformation through charcoal burning in the Caribbean), they all relate to losses of forest cover, biodiversity, soil fertility or wetlands that are widely repeated around the world to substract significant fractions of the net worldwide stocks of the resources affected.

As it is especially obvious from cases such as Laos (chapter 3), Cape Town (chapter 4), Southeast Brazil (chapter 8) and southwest Dominican Republic (chapter 10), the environmental changes documented there have direct connections to systemic changes as well, for example, with deforestation releasing carbon dioxide or inducing land cover changes in albedo and with large-scale informal housing at the periphery of metropolitan areas multiplying albedo changes and gas emissions.

However, whether linked or not, the cases of cumulative environmental change 'in themselves may pose threats to the resources on which the habitability of an ever more populous and more affluent globe will depend' (Meyer and Turner, 1995). To conclude with, a further point has to be made here that cumulative changes not only are better documented, thus providing extremely valuable material for the socially oriented exploration of environmental change, but are increasingly considered to be more significant for human activities, too (Meyer and Turner, 1995).

Regional environmental change   To understand the natural and social patterns and dynamics involved in GEC, it is increasingly recognized that global as well as regional assessments have to be carried out. Adopting the notion of 'earth transformed' and processes continuing to do so in future, it was stated by Meyer and Turner (1995) that 'these changes will find different expression and have different consequences in different regions, but they will be significant almost everywhere'. Kasperson (1992) even states that 'to understand global environmental change, regional analyses are essential'. This is, for example, most obvious from the global pattern of increased deforestation and cropland expansion which is not evident in northern hemisphere countries of the developed world where much cleared land has reverted to tree cover and land may be a net absorber of carbon from the atmosphere. Meyer and Turner (1992) point out that most changes at a global aggregate level correlate with variables ascribing human impact on the environment to population, per capita usage of resources and technology, while at a regional (and even sub-global) scale it is demonstrated by case studies that other factors are important such as 'institutions, policy and political structure, trade relations, beliefs and attitudes'. A regional valuation as such is again seen to demonstrate the need for a political economy approach in the analysis of environmental change.

It is furthermore widely acknowledged that climate change puts more stress and risk upon people and croplands near the equator so that GEC impacts upon economy and food security may be modest globally but severe for the tropical world where cumulative changes degrading land resources are most severe in any case (Meyer and Turner, 1995; Manshard, 1998 a). Again, this stresses the need for a social structure of under-standing environmental change as it occurs in the developing world.

Regional case studies of the social dimensions of GEC are required to disentangle at finer scales what is essential in different regions and what is (not) shared between the regional and (sub)global scale. Approaches as hitherto done could be classified in that either single facets of change (e.g., trajectories of deforestation or cropland expansion) are comparatively assessed or more holistic assessments attained. The methodologies to do so are diverse, and the specific value of the case studies collected in this volume is seen not to rely upon secondary and highly unreliable 'official' figures but more so to directly address the level of rural or urban land managers and also get a picture of their perception of regional environmental change and their social positioning in this.

In a broader sense, various attempts to 'regionalize' the phenomena of environmental change could be mentioned here. On the basis of sources mainly dated from the 1970s and 1980s, Hannah, Lohse, Hutchinson, Carr and Lankerani (1994) have developed mapping criteria to arrive at the classification of biogeographic realms and provinces as 'undisturbed', 'partially disturbed' and 'human dominated' zones, while the human impact here is clearly that of 'domination' and 'disturbance'. Turner, Clark, Kates, Richards, Mathews and Meyer (1990), by drawing on an inventory which covers the past three centuries, not only traced and quantified global trajectories of the human-induced transformation of environmental components but also shed light on regionally different 'driving forces' of these kinds of change. In an attempt to reconstruct spatially explict changes in global cropland cover from 1700 to 1992 – but also allowing for regional interpretations – , Ramankutty and Foley (1998, 1999) combine satellite data, socio-economic data and simple land-cover change models. From producing a global map, it is found that the results are consistent with our knowledge of agricultural geography and that (simulated) historical changes in croplands are very much in agreement with the history of human settlements and patterns of economic development. From the latter attempt, though social in a broader sense, one could not help reminding Noble (1996) who stated that much of the 'effort in ... landscape ecology is spent dealing with the technology, the challenges and the aesthetics of remote sensing and GIS', while obviously 'they are not the limiting step in achieving better landscape management and utilization'.

An example of a holistic assessment of regional environmental change is the Project on Critical Environmental Zones of the United Nations University (UNU). Drawing upon nine regional case studies and a refined concept of criticality which allows to distinguish four categories (criticality, endangerment, impoverishment and sustainability), the project focuses upon cumulative long-term changes and addresses 'regional dynamics' that are shaping the trajectories of change within each 'region at risk' (Kasperson, Kasperson and Turner, 1995). In order to approach in qualitative terms a hypothetical threshold zone of criticality, verbal as well as graphical configurations are conceived which capture the configuration of the ability of the environment to recover to its former state or capacity and the ability of society to sustain the costs of substitutes or mitigation ('nature/society condition'). The UNU and DFG/SPP task force case studies in general will not overlap except for some partial congruence in chapter 2 (Nepal) which adopts the areas's overall ranking as 'endangered' and further explores the vulnerability and livelihood security of upland farmers.

Drawing upon the UN Earth Summit's outcome in Agenda 21 of an urgently required 'earth system management' (Quarry, 1992), the modelling attempt of the Potsdam Institute for Climate Impact Research (PIK) in the form of 'Earth System Analysis' (ESA) tries to hold the balance between global aspiration and regional differentiation, between detailed knowledge and generalization, and between quantification and qualitative description (Schellnhuber and Wenzel, 1998). In doing so, it constitutes another holistic assessment that 'makes use of the simulation of more or less sophisticated copies of the planet in the laboratory of a virtual reality and owes its existence largely to the advent of electronic computers'. What makes the approach, however, different from other designs of 'regional dynamics' used is the basic idea that GEC phenomena 'should not be divided into regions, sectors or processes but be understood as a *co-evolution of dynamic partial patterns* of unmistakable character' (Schellnhuber, Block, Cassel-Gintz, Kropp, Lammel, Lass, Lienenkamp, Loose, Lüdeke, Moldenhauer, Petschel-Held, Plöchl and Reusswig, 1997). With the patterns called 'syndromes of change', some of them such as the 'Sahel Syndrome' or the 'Katanga Syndrome', thus, not neccesarily relate to the specific spatial context as implied by the use of geographical notions but more so mean a simply flowing together of many factors signifying sort of 'misdevelopments in the recent history of civilization-nature relations' that could be identified not only in the Sahel or Katanga region but anywhere else. The positioning of the case studies within specific syndromes are explored later in more detail as it is with on-going modelling attempts by ESA to which the case studies are exposed to in the concluding chapter.

Human Dimensions of (Global) Environmental Change

Climate-related research had so far been the pioneering frontier where social scientists have been included in the assessment of global environmental change (GEC). Following the Framework Convention on Climate Change in 1992 and further discussions at the Berlin Conference in 1995, particular calls to the social science community emerged 'for an improved understanding of the socio-economic consequences of climatic change and for further investigations of the response mechanisms to mitigate such changes' (Manshard, 1998 a). Assessments of GEC such as those of the Intergovernmental Panel on Climate Change (IPCC) attained to figure out - among others – who will be 'winners' and 'losers' resulting from ongoing climatic change (IPCC, 1990), thus more or less leaving social research the job of telling natural scientists how people are likely to respond to particular environmental changes and initiatives and in what way they might be persuaded to change their behavior to produce outcomes desired by policy makers. However, it was made clear by Redclift and Benton (1994), that social theory has more to offer than just designing conceptions of how the world should be and treating people in an instrumental fashion.

Some of the conceptual designs of the GEC related social science community are presented as follows, while it should be recognized that 'the community has still to develop a convincing concept for an integrative research strategy and the respective organisational infrastructure' (Ehlers and Krafft, 1998), and this not only in Germany.

The statements of IPCC relate to the highly assumed event that climatic change will mainly affect agriculture – not so much in industrialized countries where agriculture forms only a small part of the economy, but more so – in countries of the developing world where it is essential for human survival. Any projecting of global warming, for example, points to the fact that benefits could mainly be allocated to some regions in the northern hemisphere, while most of the 'loosers' will probably be found in developing nations of the southern hemisphere (IPCC, 1990; Manshard, 1998 a). Providing arguments for the particular focus of DFG/SPP's task force 5 upon the developing world, but also drawing on the main disciplinary affiliation of researchers involved here, social science research on GEC in the field of German human geography (and anthropology) is devoted a separate chapter.

We call it the framework of political ecology not because of having identified a unified human-enviromental theory everyone adheres to but more so to present an emerging environmentally oriented research agenda of what Bryant (1992, 1997) specifically called 'Third-World studies'. It is seen that political ecology can natural scientists tell more than just how people in the south are likely to respond to GEC and how they should adapt their behavior.

The Concept of Social Drivers

The definition of specific driving forces of environmental change is not always clear. In the form of 'some reasonably reliable generalizations ... about the extent and trends of human-induced environmental change', it was stated by Meyer and Turner (1995, p. 5-6) on the impact of human driving forces that

> first, human activities are comparable to or greater than natural forces as drivers of many kinds of change. Secondly, most of them have only recently become so. Yet, thirdly, acceleration and intensification of human impact, though frequently and perhaps usually the case, has not always been so. Finally, human impacts have steadily expanded in variety and character, from involving mainly some of the landscape resources of the earth – forests, soils, water, biota – to affecting the material and energy flows of the biosphere.

From the viewpoint of eco-restructuring, 'human driving forces, especially industrial development, (are seen to) have an impact on environment, especially on natural resource depletion and land-use and land-cover transformation' (Manshard, 1998 a). And similarly, from the viewpoint of ESA (Schellnhuber, Block, Cassel-Gintz, Kropp, Lammel, Lass, Lienenkamp, Loose, Lüdeke, Moldenhauer, Petschel-Held, Plöchl and Reusswig, 1997, p. 19), it was put forward that earth transformation since about 40 to 50 years could be seen as the result of a 'mega-process' that is about to bring to an end 'in a final and all-embracing manner' what began with

> the triumphal march of the bourgeois-industrial revolution which started in England some two hundred years ago (...) The ultimate driving forces of change have been natural sciences and fossil fuels, that is, an explosive combination of rather disparate gifts to humankind by history.

In an attempt to give a format to the drivers of earth transformation, in a paper commissioned by the US National Science Foundation five groups of 'social (or human) driving forces' are identified and explained in more detail by Stern, Young and Druckman (1992). The authors look upon drivers as 'a complex of social, political, economic, technological and cultural variables, sometimes referred to as (social) driving forces', given here in the order of increasing importance (Stern, Young and Druckman, 1992, p. 44-54, 76-92).

Population growth   As a matter of fact, common-sense empiricism tells that each person exerts certain demand on the environment for the essentials of life such as food, shelter and water. Demand increases with a growing number of persons what has triggered off quite many theoretical, ideological and conceptual reasoning with much of it following a Malthusian line of discourse. However, only if related to other and more important drivers, the Malthusian concept of inherent and damaging aspects of ('unchecked') population increases upon natural resources gains a special momentum.

Economic growth   First time in human history, economic activities are valued to have a global impact on the total of the earth system. Adopting here, in particular, the notion of environmental change in tropical countries, consensus exists that significant transformations of the earth in terms of land-use and land-cover changes are to be related to 'the development and expansion of western European economic orders with "capitalism" lying at the heart of it' (Manshard and Mäckel, 1995).

Technological change   Manyfold impacts of technology upon environmental change can be identified in that new ways are opened to exploit natural resources, to change the volume of resources required and the amount or kind of wastes produced per unit of output. One might recall here Lewis Mumford's (1977) conceptualization of a 'mega machine' in operation which includes cultural, technological as well as institutional (power) aspects.

Political-economic institutions   Changes of the natural environment are seen to be related and respond to actions rooted in the sphere of markets, governments and the international political economy. It could be added that though there is not yet a unified theory of political ecology, an important platform of exploring the dimensions, scale and power of the 'politicised environment' (Bryant and Bailey, 1997) in developing countries is seen here.

Attitudes and beliefs   With regard to material possession and the shaping of nature/society conditions, values, attitudes and beliefs are recognized to lie at the root of impacting upon the environment. Though not in full concordance with (or even recognized by) IHDP and IDGC, from the viewpoint of social ecology it was brought forward by Bahro (1989) that European cosmology, capital dynamics and the industrial system are part of a logical ('psycho-dynamical') chain of explanatory factors accounting for the pressure put upon the 'conditio humana' to ravage the earth.

On the concept of human drivers, it was pointed out by Spada and Scheuermann (1998) that 'single-factor explanations (...) tend to be misleading (and) the driving forces ... generally act in combination with each other (with) interactions (being) contingent on place, time and level of analysis'.

The Concept of Human Responses

Again drawing on the format as given by the US National Science Foundation to the human responses to GEC, they are seen to occur mainly in seven 'interacting systems' (Stern, Young and Druckman, 1992, p. 4-6).

Individual perception, judgement and action   The focus here is on the inputs by individuals since (aggregated) individual actions often lead to major effects, and, if organised, can influence collective and political responses.

Markets   Markets are important in that GEC is likely to affect the prices of important commodities and factors of economic production. So far, however, existing markets seemedly do not provide the right price signals for managing global environmental change.

Socio-cultural systems   Held together by bonds such as solidarity, obligation and duty, socio-cultural systems may develop ways of interacting with the environment which are held suitable social strategies of effective response (even as informal ties).

Organised responses at the subnational level   They are esteemed to be important in their own and by influencing the adoption and implementation of government policies.

National policies   Not only in the form of environmental policy, but also including economic, fiscal, agricultural, forestry, science and technology oriented policies, they aim at coming to international agreements and affect the ability to respond at local and individual levels.

International co-operation   The formation of international institutions to respond to GEC is considered to be the key factor in solving problems. International co-operation, including both governments and non-state actors, already proved to be important in addressing systemic GEC such as ozone depletion and global warming.

Global social change   It is seen to influence the way humanity will respond to the prospect of GEC and its ability to adapt to such changes.

The social or human drivers of GEC as well as the systems of response constitute major human proximate causes or response systems which either directly alter aspects of the natural environment or assumedly mitigate effects of earth transformation. It has been emphasized by Stern, Young and Druckman (1992) that outlining drivers and responses as such has 'only heuristic, not explanatory value'. Nonetheless, it is somehow ironic that the question of what actually constitutes 'social' is not addressed at all. As a consequence, the social perspectives of GEC – comparable to those of 'desertification' – tend to 'degenerate into a pluralist grab-bag of ideas, embracing everything from land tenure to international organizations' (Watts, 1985). By putting it this way, we do not suggest, for example, that the drivers of GEC or the claimed 'misfit' between human and ecological subsystems have to be incorporated into a holistic and watertight social theory, but the point has to be made that any serious discussion of social relations must begin with a notion of structure. According to our understanding, the newly emerging field of political ecology has to offer not only characteristic combinations and interactions of variables, incorporating both drivers and responses specifically worked out in the human-environmental context of the developing world, but more so stresses the notion of 'social' implied.

The Framework of Political Ecology

Definitions, Topics and Social Strategies Involved

While there is growing pioneering work in political ecology since the late 1980s, the definition of it is not always clear. As many, especially anglo-american scientists, have noted, Wolf (1972) was the first to employ the term, but the usage of it did not have the momentous impact of the seminal work of Blaikie and Brookfield (1987) in which they developed a conceptual framework of 'regional political ecology'. However, and drawing this information from Ante (1985), the point has to be made that in an attempt to combine specific aspects of the natural environment (*Landesnatur*) with political structures and socio-economic settings, the concept of *politische Ökologie* had earlier been employed by Heberle (1945, 1978).

In combining here the perspectives from cultural ecology and political economy which have been prominent in anthropology and geography (Hardesty, 1977; Netting, 1986; Peet and Thrift, 1989; Emel and Peet, 1989), the limitedness of traditional cultural-ecological studies due to their focus on the micro-scale (i.e., to explore the intricate, complex interactions between peoples and their environments in the context of resource use) was compensated by integrating the role of political economy (i.e., to focus upon the nature and significance of the unequal distribution of power and wealth in society). Thus, political ecology became characterized by the notion of 'difference' in terms of a social structure pervading the analysis (Johnston, Taylor and Watts, 1995 b; Harvey, 1996) what is increasingly done along the lines of 'agency/structure' (Giddens, 1984, 1993). Though a political ecology of much broader content than given here underlies the definition as given by Keil, Bell, Penz and Fawcett (1998, p. 1),[2] their words seem useful to shed light on the present and general state of conceptual development.

> Political ecology ... at present ... raises more questions than it answers. But these are timely, and in some cases unique questions. Attempts to come to grips with the environmental crisis have opened up previously unseen landscapes, and the theorizing has begun.

It had early been noted by Enzensberger (1973, 1974) – thus reflecting much of the current concerns and ambiguities of integrating social and natural sciences into GEC research – that with political ecology a 'hybrid' discipline had been created which uses natural as well as social science categories, methods and tools while the value added by doing so is theoretically not made explicit. Hard (1997) went as far as to state that the 'hybrid paradigm' created by political (or 'symbolic') ecology, on the one hand, is an ecology of a much desired broader type, while on the other hand, and in structural terms, it is nothing more than just similar to the concerns of classical geography and traditional vegetation studies.

Despite of these warnings, the work particularly of Blaikie and Brookfield (1987) turned out to be a propulsive force driving a plethora of studies using the term. Notwithstanding the fundamental truths of both Enzensberger's criticism and Hard's irony, political ecology is seen to constitute a fruitful field for further explorations in particular if reference is made to clarify the much claimed 'misfit' and not to forget about a notion of social structure. Thus, it is seen here to constitute a suitable platform for the social analysis of livelihood systems in endangered ecosystems of the developing world by contributing to the social conceptualization of GEC and, in the best of all cases, integrating the outcomes of natural science research.

In the context of understanding the human dimensions of global environmental change, a first mention was given by Stern, Young and Druckman (1992) in that main findings were seen to be expected mainly from 'marginal' (or 'hybrid') fields such as political ecology. It had further been widely demonstrated how the social perspectives of regional environmental change in West Africa are successfully dealth with by political ecology – e.g., Bassett (1988), Geist (1992) and Krings (1994) among many others – as compared to the theoretical untidyness of traditional or 'orthodox' studies simply providing inventories of anthropogenic forces central to the modification of Sahelian ecosystems.

In terms of a first conceptualization as a research perspective of GEC, we see political ecology to have been introduced and employed by Watts and Bohle (1993 a) and Bohle, Downing and Watts (1994). It was later given a mention by Spada and Scheuermann (1998)[3] mainly drawing from Bohle's input in the form of an outline of the research perspectives of German (human) geography and ethnology. In an attempt to value the spectre and potentiality of political ecology emerging as a platform of GEC research, some of the main topics of 'Third-World studies' are summarized as follows.

Global/local interplays of earth transformation   In combining the two perspectives of cultural ecology and political economy, it is made explicit that human-environmental relations can only be understood at local, regional as well as global scales by examining the relationship of patterns of resource use to political-economic forces. While one of the major intellectual developments in social science to grapple with phenomena of 'globalization' has been a better understanding of space and time, the 'problematizing of both (...) repositioned a previously marginal geography toward the center of the realm of social sciences' (Taylor, Watts and Johnston, 1995).

   Here, a first common element of political ecology models is the progressive contextualisation of human-evironmental relationships at different (spatial) scales of enquiry. While geographical scales span in their extent from 'a single point to the entire globe' (Meentemeyer, 1989), the notion of political ecology is the understanding that the concrete spatial context (the 'real' milieu, environment or location) actually is 'placeless, for the reality within which we live is the global world-economy' (Johnston, 1986). That the functional imperatives of the global capitalist economy, however, do not unilaterally impact upon the local scale but are broken down by complex interactions in specific socio-cultural and political contexts modifying and even neutralizing various actors' agency potentiality has become more and more recognized (Schneider and Geist, 1996). Located between the micro-level or 'scale of experience' (following the logics of local/regional livelihood systems or modes of production) and the macro-level of the world-economy (driven by the logics of capital dynamics) is the meso-level or 'scale of ideology' (Taylor, 1981) which relates to the national state that follows the logics of power and is heavily adherent to world market conditions.

   The use of local, regional and global scales just relates to 'conceptual levels' while more types of scales (such as temporal, functional, absolute and relative) will be involved and thus have to be considered (Harvey,

1969; Meyer, Derek, Turner and McDowell, 1992; Gibson, Ostrom and Ahn, 1998). By bringing, for example, time and space together, it is suggested by Taylor, Watts and Johnston (1995, p. 8-9) not to pursue 'the familiar analogy between the two, such as equating time periodization with space regionalization', but to privilege three spatial scales (as introduced above) and three time spans, i.e., the long-term perspective (or Fernand Braudel's *longue durée*), the short-term view of (multiple) eventism, and the medium-term span of cyclical changes (such as 'hegemonic' or Kondratieff cycles). It is noted that

> (a)s with spans, so with scales a sensitive analysis does not argue for one against another but focuses upon their relations. Local communities may be buffeted by global forces but they are not helpless victims with no coping strategies. However, neither can they be autonomous of the world they inhabit, so that their strategies will invariably involve consequences beyond their direct control. In this case geographers deal with a local-global dialectic, where local events constitute global structures which then impinge on local events in an iterative continuum.

In an attempt to identify variables or 'causal factors' interacting between the scales, it was suggested – as some of the outcomes of the causal structure analysis of famine – to distinguish between several types of causes (or risks) (Bassett, 1988; Cannon, 1991; Bohle, Downing and Watts, 1993; Watts and Bohle, 1993). First, 'ultimate causes' or 'initital conditions' (such as unequal entitlements and surplus extraction by a non-producing class) are structural or systemic in nature and often create 'predisposing conditions' of human-environment interaction. Secondly, 'proximate causes' relate to the more immediate and 'situational' forces (such as indebtedness). In the form of 'trigger events' or 'stressors' (such as drought or high prices of grain), the latter are often catalytic factors leading to distress or the collaps of land-use patterns. That such a view has to be exposed to varying modes of perception by different actors has become one of the newly emerging strands of political ecology dealing with the socially constructed concepts of 'nature', environment and environmental change (Blaikie, 1994, 1995; Biot, Blaikie, Jackson and Palmer-Jones, 1995; Harvey, 1996). It is also one of the major fields in chapter 7 of the volume illustrating the partly overlapping, partly conflicting various environmental perceptions of drought in rural Botswana, and in chapter 9 contesting indigenous views upon the environment and the developmental perspective as molded according to the Western style.

To conclude with, the increasing interest in global phenomena has not only shifted geographic studies more towards the meso- and macro-scale, but political ecology (and human geography) is about to model global/local interplays the specific web of which allows for rewarding interregional and intercultural comparisons.

Starting, for example, from the formulation that GEC is a result of changing modes of production and consumption, two lines of arguments – partly conflicting and partly supporting each other – could be identified and have to be tested against varying regional and cultural contexts. First, the global expansion of markets increases robustness and flexibility of decision-makers in the economy and will contribute positively to meet basic needs and sustain the environment. Secondly, local socio-cultural systems function as security networks for individuals to guarantee life chances and survival in the context of natural environments not getting 'mined'. From a large body of case studies, one could draw some preliminary conclusions with regard to regions and processes, for instance, as different as land degradation in west african savannah areas and deforestation in the Amazon basin. While in the first case environmentally threatened modes of subsistence production in rural areas of West Africa tend to be mitigated by chances of urban wage labour through a network of social relations, in the case of Amazonia, wealthy economic agents following profit signals sent from the market are about to marginalise people though they have developed flexible socio-cultural response systems with the overall result that the latter lose their land and have no access to urban wage labour (Stern, Young and Druckman, 1992).

All of the case studies collected in the volume will contribute to refining specific webs of global/local interplays in one way or another. While there was (and is) tremendous work on the Sudano-sahelian zone of Africa as well as on deforestation in Amazonia, the case studies, however, will contribute to the knowledge of hitherto not so much covered ecozones such as coastal forest ecosystems, dry forest zones and mountain ecosystems by putting there the livelihood systems of not so wealthy people first.

Land-use and land-cover changes   Williams (1994 a, p. 97) noted that the felling of trees for the combined objectives of obtaining wood for construction, shelter and toolmaking, of providing fuel for commercial as well as domestic purposes, and of creating land for agriculture has culminated in one of the main processes whereby humankind has modified the world's surface cover of vegetation. Despite the importance and magnitude of this process, the distribution, quantitative extent, and rate of change in the area of forest, through both deforestation and reforestation, have been and remain subjects of great debate and uncertainty.

Many of the case studies collected in the volume, be it the case of Laos, Tanzania, Malawi, Brazil or that of the Dominican Republic, will enter the debate on deforestation and probably increase 'uncertainty' by pointing out the specific socio-cultural and political-economy context in which tree felling occurs.

Here, a second common element of political ecology models becomes evident. Against the background that complex interactions between environment and society are always embedded in history and specific ecologies, it is important to understand such processes in their historical positioning with a contextual analysis emphasizing the transformation of indigenous systems of resource management in the course of their incorporation into the global economy. An arena of particular interest for studies using the essence of this approach (and/or not using the term political ecology) was the impact of colonial policies on human-environment relations (Franke and Chasin, 1980; Watts, 1984; Geist, 1996), while other interests included environmental history and/or the role of the state in deforestation (Hecht, 1985; Krings, 1996). This can even be called a third common element of political ecology models, i.e., an emphasis of the influence of state intervention in rural economies on land-use patterns.

The second and third elements inherent in political ecology are seen to be derived from or sometimes even explicitly based on Ernst Friedrich's (1904) concept of *Raubwirtschaft* or *Raubbau* ('destructive exploitation'). The concept allows for distinctions between natural resource exploitation having no societal consequences (*einfache Raubwirtschaft*) and such calling for human responses (*charakterisierte Raubwirtschaft*) and has triggered off a large body of comparative (but partly 'romanticized') frontier or colonization studies in the United States (less so in Germany). These or similar themes are seen by Williams (1994 b) to have culminated in the opportunity to plan the 'Man's role in changing the face of the Earth' symposium in 1956 with direct successors such as the symposium that produced 'The Earth transformed' (Turner, Clark, Kates, Richards, Mathews and Meyer, 1990). The notion arose that much of the damaging impact upon natural resources in the developing world started with the 'Columbian encounter'

(Turner and Butzer 1992), while Williams (1994 b) clarified that 'the transformation of the earth has occurred for many reasons, but two are paramount: the explosive increase of European population and its movement overseas, and the rise of the modern capitalist economy and its evolution into industrialism'.

The initial focus focus in political ecology highlighted the relations among state policy, surplus extraction, accumulation and environmental degradation, thus providing a necessary corrective to viewpoints asserting that environmental problems are mainly related to Mathusian-like pressures, peasant irrationality and ignorance. Another strand of research started to explore issues relevant to the development literature, i.e., the rural labour question and other key aspects of institutional forms of the penetration of the production process by capital and the state and how this links peoples and environments. While in some cases concepts of natural science such as entropy and thermodynamics were applied and linked to the classical political economy of the industrial sector (Altvater, 1998), in other cases the rural political economy of Karl Kautsky (1899/1966, 1927, 1988) was seen a fruitful platform of theorizing (Watts, 1989, 1990, 1996). In particular, the issues of agrobusiness and contract farming were raised (Little and Watts, 1994; Grossman, 1998). However, a tendency in political ecology research became evident that theorizing could easily get trapped by focussing on the effects of political-economic forces on the natural environment in a unilinear fashion. Zimmerer (1996) went as far as to state that even 'most political ecology has conceived the environment solely as a receptor for modification'.

At this point, the origins and common elements of political ecology might be remembered in that the initial starting point of analysing human-environmental relations is at the local level, i.e., examining resource-use patterns as they relate to households as units of production, (gender-based) struggles within households, or relations and conflicts among households. This was summarized by Blaikie and Brookfield (1987, p. 27) as follows:

> It starts with the land managers and their direct relations with the land (crop rotations, fuelwood use, stocking densities, capital investments and so on). Then the next link concerns their relations with each other, other land users, and groups in the wider society who affect them in any way, which in turn determines land management. The state and the world economy constitute the last links in the chain.

Here, a fourth common element of political ecology is evident, i.e., the focus at the local level on differential responses of decision-making units to changing social relations of production and exchange. In linking the level of rural or urban households to the wider and international ('global') context, increasing efforts are spent in formatting the issue how functional imperatives of the world economy – or (dis)order – translate into local-level milieux. From the viewpoint of 'rural' political ecology, two of them are provided here.

First, a 'chain of explanation' is provided by Blaikie (1994) and, though this relates to land degradation, the linkage to environmental change is immediate. He suggests several modes to contextualize how 'site', 'symptom' (of degradation), (land use) 'practice', 'decision-making', 'society', 'state' and the 'world' could be linked under varying contexts of analysis.

Secondly, specifying the ravaging of the earth in terms of nature/capital conditions – what also applies to societies having (had) run a namedly socialist type of economy – , Rauch (1996) explores the regional differentiations of technological development and expanding capital and how the logics of accumulation penetrates rural areas and societies in the developing world in a manner not sensitive at all to the ecosystem properties given. The fully disproportionate relation between ecosystems and the world market is seen to be reflected at the meso-level when the interests of ruling class members (*Staatsmacht, Staatsklasse*) in natural resources are explored (Rauch, 1996, p. 73).

> From a tendency to privilege themselves and reacting upon pressures for legitimization, (ruling class members) generally employ short-term strategies of capitalizing the natural environment. Very often, natural resources constitute the material basis from which rents can be drawn and projects financed in order to legitimate their power (...) The rent-seeking behaviour could be seen to go hand in hand with a tendency to nationalize the control over natural resources, thus robbing traditional authorities any control capability and henceforth responsibility directed towards the regulated and sustainable use of resources. However, members of the *Staatsmacht* very often are not able – or not willing for the sake of self-privileges and appropriation – to exert responsibity in an effective manner. This inevitably reinforces any tendency towards uncontrolled *Raubbau*.

The tendency of capital to set in value those resources, areas and societies where conditions allow for comparative advantages in the production of primary products and exploitation of ecosystem properties is seen to create two types of regional economies, i.e., *Boom-Regionen* (boom regions) directly responding to specific demands of the world market and *peripherisierte Regionen* (marginal regions) being excluded from world market demands at present. Both types are seen to constitute only the extreme ends along a continual scale where positional shifts between the two are occurring frequently. By linking to the spatial classification the properties of resource use and modes of (smallholder) production, the framework could be recognized as a valuable platform from which to categorize the global/local interplay in rural areas of the developing world – see Table 1.2.

Table 1.2   Patterns of regional rural development in relation to resource usage and world market demand

World market's demand for regional resources	Mineral Resources	Agricultural Resources	
		Rich	Poor (degraded)
Strong		*Boom Regions*	
		(externally dependent, crisis-prone monostructure)	
	Negligence of rural population needs; weak social position	Exploitation of rural population; relatively strong social position	Dual economy selective usages; high input of external resources
	Mining and oil-producing regions	Export-producing peasantry in West Africa	Semi-arid regions with irrigated export produce
Weak		*Marginal Regions*	
		(dominant subsistence production)	
	Crisis-ridden since main source of income lost	Main problems: market access & land reform	Main problems: land use & population pressure
	Zambia	(Africa, Asia & Latin America)	Mountain areas (Nepal, N-Ethopia)

Source: Adopted from Rauch (1996), *Ländliche Regionalentwicklung*, p. 75.[4]

In an attempt, to rank the case studies in the volume according to the outline as given in Table 1.2, one will identify a most diverse pattern of factor combinations such as (i) weak demands from the world market as found in marginal mountain areas of Nepal (chapter 2) and the Dominican Republic (chapter 10), (ii) strong demands from the global tobacco market as in highlands areas of Tanzania and Malawi shifting from rich to poor resource endowments (chapter 5), (iii) most recently occurring shifts from weak to strong world market demands as in the case of seed bean production invading traditional Maasai grazing areas in Northern Tanzania (chapter 6) and in the case of an all-embracing vortex of developmental activities mainly originating from hydropower investments in rural and hitherto weakly integrated areas of Laos (chapter 3), and (iv) first advances from the side of (non-)governmental agencies to shift an Andean marginal and resource-poor pastoral area closer to the demands of the world market and Western style development (chapter 9). While it is not attained here, to further format the cases according to a watertight theory of global/local interplays, each individual case will take the argument much further.

Common/private resources and environment   Blaikie and Brookfield (1987) noted that 'the discovery by social scientists that vast numbers of farmers, pastoralists and fishermen have been managing 'common property resources' (CPRs) in pursuing their livelihoods is akin (...) (to give a) new name for an old phenomenon, and we realize in the case of CPRs that we have been involved with it all along'. However, Blaikie and Brookfield's (1987) notion that 'where CPRs are encroached upon and privatized through enclusore, the remaining areas have to carry the added displaced load of the CPR users', is not always properly understood. The issue, expressed also in related terms such as 'public goods', 'open access', 'collective action' and 'appropriation', is at the heart of the GEC research agenda and recent reviews on this are given – among others – by Ernst (1998) and Gibson, Ostrom and Ahn (1998).

Some writers have tried to show that socially agreed upon decision-making rules will inherently tend to break down and bring about the destruction of the CPR. Essentially restating the Malthusian dilemma, the formulation done by Hardin (1968, 1970, 1993), i.e., the 'tragedy of the commons', is the most celebrated and the most widely criticized. The persuasiveness of Hardin's argument (the pursuit of short-term gain will demonstrate the inefficiency of CPR arrangements inevitably, i.e., in the sense of a 'solemnity of the remorseless working of things') has led many to urge privatization of CPRs. However, from a growing body of political ecology research on 'Third-world' situations it could be concluded that 'formalized CPR arrangements are not a quaint anachronism inherited from a pre-industrial past, but can have positive and enduring benefits to all users' (Blaikie and Brookfield, 1987).

As a matter of fact, CPRs and private resources (PRs) frequently have very close relationships, and pressures upon one set may well be transmitted to the other (Ernst, 1998; Geist, 1998; Lohnert, 1998). Several exemplifications, however, lead to the assumption that any conceptualization of how to reduce deleterious pressure upon natural resources in the context of wealth versus poverty and ownership versus landlessness should take into account that there might be positive reactions especially among individuals and groups in the more affluent countries, but 'the dynamo of capitalism is against them (...): as a whole, if not as individuals, we are pressed to act against our long-term interest in order to maintain our short-term position' (Johnston, Taylor and Watts, 1995 a). To conclude with, several statements on CPRs are given illustrating the significance for a research agenda on 'Third-world'-related GEC – with most of them provided by political ecology.

First, the UNU project revealed that any of the highland studies settled in developing countries 'suggest the increasing vulnerability of regional populations that may accompany the transformation from traditional common-property arrangements to privatization and commercialization' (while only the case of the North Sea is 'encouraging') (Turner, Kasperson, Kasperson, Dow and Meyer, 1995).

Secondly, it is made clear by Bryant and Bailey (1997, p. 163-4) that not a 'tragedy of the commons', but a 'tragedy of enclosure (of the commons)' has to be put on the agenda and the implications for marginalized grassroots actors to be discussed.

> Indeed, research by political ecologists has been instrumental in pointing out that the Third World's environmental crisis reflects mainly a tragedy of enclosure rather than a tragedy of the commons (...) In this process, the state, often acting in conjunction with businesses and multilateral institutions, denies grassroots actors access to commons resources hitherto managed by them through local institutions such as CPRs. In effect, CPRs are taken over by the state for large-scale commercial exploitation either by its own agencies or by allied business interests using the legal-political powers of the state. A notable case in point has been the creation of extensive networks of reserved forests, national parks and 'government lands' in many parts of the Third World (...) The habitually exploitative practices – styled 'development' in the postcolonial era – carried out in these 'nationalised' territories has included notably large-scale logging, mining, cattle ranching, cash-crop production and dam construction (...) / The first thing to note in this regard is that (...) a corollary of

'development' has been undoubtedly the weakening, if not the elimination altogether, of grassroots environmental management in much of the Third World. / The second point to note is that the enclosure of the commons served to further marginalise poor grassroots actors in the measure that their access to environmental resources essential for their livelihoods was restricted or denied. Not only was access to commons resources ended, but these actors were often forced into a situation whereby they had to work ecologically marginal lands elsewhere in order to survive (...) The end result was that marginalisation became a defining trait for most ... as they were displaced from newly created reserved forests, national parks or other 'development' projects (.../...) Yet to speak of the marginalisation of poor grassroots actors is also to acknowledge that some actors within this broad category have been worse affected than others ... Poor women and indigenous minorities in particular have apparently borne a disproportionate share of the costs with such marginality.

Be it the denied land access in valley bottom areas in the southwest of the Dominican Republic or in northern Laos, be it landlessness in Brazil or informal squatting in urban South Africa, all of the chapters in the volume will address the issue of the commons either implicitly or explicitly stated, and mostly conceptualized in terms of land tenure and infrastructural impacts of what commonly is styled 'development'. In cases where no commons exist, the general point has to be raised here (and will be obvious from chapters 3 and 9 in particuar) that 'indigenous knowledge usually reflects a detailed appreciation and understanding of local environmental resources by grassroots actors, and that such knowledge has often served as the basis for highly effective environmental management systems allowing for simultaneous resource exploitation and conservation' (Bryant and Bailey, 1997).

GEC, conflict, social justice and uneven development   Different from the driver/response concept, the (natural) growth of population with its consequences for food production and distribution is seen to be not the only population characteristic generating or driving global change. Drawing on the notion of 'people in turmoil' (Johnston, Taylor and Watts, 1995 b), a difference could be made up between behavioral (i.e., direct, intentional and physcial) violence with resulting population 'decreases' from political and military strife, and even larger impacts upon populations in the form of what is known as structural (i.e., indirect or 'silent') violence. The inputs from peace and conflict analysis are seen to be so far not well represented on the GEC agenda of human drivers and responses. There is no sensitivity to the map of uneven development underlying, for example, unequal life expectancies which involve structural violence in the developing world or the problems of hunger and survival that will continue to present daily concerns (if not crises) to many and assumedly the majority of the earth's population.

A large and growing body of research has related these problems, inherent in the class structure of the social positioning of individuals and groups in (unequal) power relations, to the issues of entitlement, vulnerability, livelihood security and environments (Johnston, Taylor and O'Loughlin, 1987; Watts and Bohle, 1993 a, b; Lohnert, 1995; Rauch, Haas and Lohnert, 1996), but also demonstrating the immediate and direct linkages to global environmental change (Bohle, Downing and Watts, 1994).

Another population characteristic not well represented on the agenda of social GEC research (in particular, IDGC's attempt to brigde the 'misfit between ecosystems and institutions') are the spatial implications of people responding to push factors impelling them away from some areas and responding to the pulls of more attractive places elsewhere. Inherent in matters of mobility are not only possible tensions (easily directed towards culturally recognizable immigrants if they are perceived as threats), but also impacts upon political movements (including the valuing of environmentalism) often fanned by xenophobic groups and directed towards refugees the number of whom is increasing due to both famine and other effects of structural violence. In particular chapter 4 enters this arena of political ecology.

Instead of, or in addition to 'looking at human systems as subsystems of the ecosphere' (Pritchard, Colding, Berkes, Svedin and Folke, 1998), the aspect of 'difference' matters in global change as put forward by Johnston, Taylor and Watts (1995, p. 150).

> Every society has norms which underpin its social relations; these include ... definitions of acceptable roles and behavior for individuals and of micro-social organizations, such as household structures. Those norms invariably sanction unequal power relations between groups within society. Such inequalities may be challenged and altered, though usually only after substantial struggle.

From a normative point of view, environmental change is neither good nor bad in itself. The nature of natural resources is simply 'neutral physical stuff' (Zimmermann, 1951), while societal interactions with the geophysical and biochemical milieux are what make those either resources (i.e., environmental improvements) or hazards (i.e., environmental threats or degradation). But even with environmental degradation, from a political ecology point of view it is the 'capability', for instance, of land to satisfy particular user needs which matters. The paradox has to be broached why and for whom the problem of environmental change arises at all.

'In most environmental changes, even ones that are more harmful than not, there are individual winners as well as losers' (Meyer and Turner, 1995). From a policy or action oriented view on human responses and the costs of mitigation, thus, matters of social (in)justice are raised. They are seen to be reflected in terms of how sustainable development is conceived and what the societal preconditions are to this.

Conceiving development so as to involve matters of social (in)justice clearly relates to any concepts aimed at the betterment both of material living conditions (e.g., food, clothing, housing and health) as well as assets, entitlements and immaterial conditions (such as social justice). While the normative (and humanistic) component of conceptualizing development as such is obvious (Seers, 1974), the linkages to sustainability are still under discussion as it is with any attempt to quantify the normative as well as material aspects. Having owned large parts of its success from its opacity, the most consensualist definition of sustainable development is that of the UN Brundtland Commission stating that this is 'development that seeks to meet the needs and aspirations of the present without compromising the ability to meet those of the future' (WCED, 1987).

Probably the small part that agriculture plays in their economy (and thus will be affected by climatic change) 'has led some economists (mainly in the United States) to conclude that climate warming is not a great concern and that major efforts to counteract it would not be economically justified' (Manshard, 1998 a). However, agriculture as a basis of human survival is essential in economies of the developing world and it could be misleading to let such positions enter the sphere of international agreements. As matter of fact, it was pointed out by Meyer and Turner (1995, p. 312) that

> (e)nvironmental-hazard studies suggest that the poor and unempowerd are the most vulnerable to these impacts because they have fewer options for adapting. Yet it is not clear that the poor and unempowered would not also disproportionately bear the costs of action taken to prevent these impacts.

At this point of exploration, it could be recognised that the absence of conclusive evidence on the nature of GEC together with or 'exacerbated by the lack of detailed knowledge of how sustainable development could be ensured given the existing pressures on the environment' not only blocks 'a major educational task' (i.e., immediate action if the pessimists' expectations are to be averted) but more so is associated with a group of problems 'which is largely political in its structure' (Taylor, Johnston and Watts, 1995 a, p. 299).

> Much of the 'rape of the earth' to date has been undertaken either by or for the populations of 'developed world' countries. With a changing world political order, the governments representing the peoples whose lands and livelihoods have been exploited in these unequal relationships argue that they should not pay the price of resource depletion, for which they have not been responsible and from which their populations have benefitted very little.

Though it may well be premature to expect consensus on this, and much less from a still to be unified theory of political ecology, some of it has been touched in the works of Harvey (1974, 1996), Blaikie (1986, 1988), Smith (1984), O'Connor (1994) and Bryant and Bailey (1997). However, in general, little is alluded to the empirically researched situations facing people living in countries on the periphery of the global capitalist system. A prolonged discussion still has to be induced of what capitalism looks like from the point of view of the majority of humans being alive today in the neighborhoods and villages of urban and rural Africa, Asia and Latin America and being exposed to changing environments (or environments at risk). Having adopted the individual household level as a basis of empirical research, most of the contributions in this volume will directly respond and contribute to this.

Chapter 2, for instance, enters both the debate of conceptualizing the various dimensions of sustainability and searches for appropriate methods for the social analysis of livelihoods in endangered ecosystems drawing from the case of Nepalese upland farmers in remote villages of the Himalayan mountains.

The notion of societies 'at risk' (Blaikie, Cannon, Davis and Wisner, 1994) as raised in chapter 2 forms also part of the case study in chapter 9 dealing with a marginal pastoral community in the Andean highlands of Argentina particularly entering the rather new field (in anthropology) of investigating institutional risks arising from the capitalization of traditional modes of production in a harsh natural environment.

In the case of Brazil, for instance, 'the confrontation between liberal environmentalism and indigenous environmental political action is exemplified in struggles over rural land as livelihood, in which the state is deeply implicated' (FitzSimmons, Glaser, Monte Mor, Pincetl and Rajan, 1994). That this reality, brought forward by a number of case studies, might in future well involve officially protected wilderness areas is at the heart of chapter 8 settled in the southeastern part of Brazil.

And what, again using the Brazilian case, was exemplified as the military government's construction of immense hydroelectric dams and the challenge to the dams mounted by displaced small farmers (McDonald, 1993), could be compared with the situation in Laos under the terms of a neoliberal restructuring of economy impacting on environment as put forward in chapter 3. While in Brazil, the farmers' effective response to their displacement has changed the discourse of state responsibility in infrastructural development, in the case of Laos a changing discourse of shifting cultivation and (village) forest management is presented as an outcome to be expected.

In chapter 4, sort of a socio-ecological trap is presented in the form of rural-urban migrants coming to Cape Town metropolitan area. They were pushed by unsustainable modes of livelihoods in their rural areas of origin, pulled by attractive job opportunities and got entangled in the contradictory and critical crisis conditions of urban ecology, insecure land tenure and the contradictions of urban labour markets. The chapter also points to the fact that most of the research on human dimensions had so far been done in non-urban settings and that a somehow different track in social GEC research has to be taken. This is not only to suggest that an expanded view of 'urban metabolism' is required in terms of a set of variables mediating between environmental hazards and human responses (Mitchell, 1998), but also to deeper understand the crisis conditions of urban livelihood systems and their interacting with environmental change given the unbroken trends of inmigration and a 'people at turmoil' (Johnston, Taylor and Watts, 1995 b).

It was pointed out by FitzSimmons, Glaser, Monte Mor, Pincetl and Rajan (1994) that the political construction of environmentalism (say, notions of GEC) in countries and democracies of the developed world 'raises a number of provoking questions about the confrontation of environmental movements with the liberal state and about the liberal form of institutional environmentalism that results'. Yet the precise form of GEC related influence in economic, political and cultural terms will probably be found to differ substantially with different geohistorical developments having lead to quite different outcomes not fully understood for a representative majority of countries in the developing world.

At this point, two remarks will be introduced, one relating to a fundamental group of problems linked to the common treatment of natural resources as private property, and the other to how environmentalism is created having reached a spectacular peak in the UN Earth Summit Conference and then declining due to the conditions of the political economy found in the developed world.

It is noted by Johnston, Taylor and Watts (1995 a, p. 299) on 'changing the changing relationships with nature' that

> much resource depletion and most environmental pollution comes about because of individual (including corporate) actions, each of which in itself is a very small, marginal contribution to the growing problems. Those actions take place within a mode of production whose political and other leaders increasingly promote the private ownership of all means of production, including nature: within capitalism, economic survival demands that resource exploiters continually increase their pressure on nature in order to sustain their competitive position in world markets.

While this relates to (inter)national economic (dis)orders to be discussed within the framework of an emerging social agenda on global environmental change (Altvater, 1998), another aspect of structural violence is seen to be obvious. Reference is made to the outlining of 'serveral possible lines of connection between First and Third World political struggles, asking how a particular liberal-capitalist political culture seeks to become hegemonic internationally and thus to define, structure, institutionalize, and constrain environmentalism' (FitzSimmons, Glaser, Monte Mor, Pincetl and Rajan, 1994). In the special context of GEC awareness and policy, for example, Schellnhuber, Block, Cassel-Gintz, Kropp, Lammel, Lass, Lienenkamp, Loose, Lüdeke, Moldenhauer, Petschel-Held, Plöchl and Reusswig (1997, p. 19) note that

> about two decades ago the sciences, the political arena and the media began to recognize the significance of ... problems (of GEC). With the fanfare announcing the discovery of the hole in the ozone layer over Antarctica, world opinion was finally shaken awake and pushed into a condition of hyperactivity ... Today, 5 years after Rio, concern about the condition of the 'patient Earth' has considerably declined, but not because the problems have actually become less acute. The reason is that, particularly in the industrialized countries, the worries about unemployment, criminality or the costs of social security systems are again generally being discussed as purely social and often national problems – essentially divorced from the global problems of the civilization-environment interface.

That the two latter remarks do not form an ideological critique remaining unconscious of its own limitations - as clearly and earlier stated by Enzensberger (1973, 1974), i.e., pointing out the bourgeois origins of GEC concerns in the sense that it was only when environmental deterioration started to foul capitalism's own backyard, and maybe threaten the process of accumulation, that much attention was paid – we will try to outline a preliminary framework or platform for the conceptualization of livelihood systems in endangered ecosystems of the developing world. Particularly emerging from the outcomes of the South African case study (chapter 4), the attempt to conceptualize societal preconditions of how to deal with 'winners' and 'loosers' of environmental change in the developing world is seen to be rooted in the difference which peace and conflict research makes between the conditions of social justice (or positively meant 'peace') and social injustice (or 'structural' and 'indirect violence' as opposed to 'direct' or 'personal violence'). Both a conceptual design of multi-dimensional sustainability, vulnerability and structural violence is attained to be linked to social GEC analysis.

Sustainability, Environment, Vulnerability and Coping Strategies

In an outline of the research perspectives in the fields of German human geography and anthropology focussing upon the developing world, Bohle depicted the relationship between GEC and sustainable livelihood security by using a diagram having five partly interacting boxes with each box showing the major field of concern and the risks to be encountered there (Spada and Scheuermann, 1998).

Among the five overall research concepts esteemed to be particularly suitable approaches for social GEC research, political ecology (together with cultural ecology) is one of them, while the others are (i) risk analysis, crisis and conflict research, (ii) vulnerability analysis and mapping, (iii) carrying capacity analysis, and (iv) energy flows and nutritional systems analysis. The five major research concerns surrounding the GEC-livelihood relation are (i) natural resource management – influenced by factors of demography and poverty and at risk through climate change, variability, criticality and uncertainty, (ii) demography – interacting with the management of natural resources and urbanization and shaped by the risk of unsustainable population growth, (iii) poverty – impacting resource management and migration and bearing the risks of endemic hunger, famine crisis and vulnerability, (iv) migration – impacted by poverty, contributing to urbanization and involving the risks of disintegration, pressure and conflict, and finally, (v) urbanization – impacted by migration and demography and bearing the risk of unsustainable metropolitan growth. While any of the case studies in the volume could be attributed to one or several of the research concerns outlined, Bohle further stressed that social GEC research should be given a format or structurally organized along the lines of 'syndromes' and 'transects'.

While the linkage to the transect approach will allow to connect social science research with gradients of global change that are chiefly determined by biophysical features, the linkage to the syndrome approach corresponds with the requirement of political ecology to begin any serious discussion of social relations (to the environment) with a notion of structure.

It had been pointed out by Watts (1985) that the social perspectives, for example, of dryland ecology (or 'desertification'), often 'degenerate into a pluralist grab-bag of ideas, embracing everything from land tenure to international political organizations'. While neither the US National Research Council's summarization of drivers and responses nor the UNU project's qualitative assessment is seen to be totally free from simply providing inventories of anthropogenic forces central to the assessment of criticality, it is not suggested that such agents must be incorporated into a holistic and watertight social theory but rather to make explict a notion of structure. In the context of 'syndromes' (as well as with most of the political ecology analysis) this is done in terms of an explanation of human action within the framework of dialectical relations between action and structure (Giddens, 1993).

Syndromes   The concept of 'syndromes of change' had earlier been introduced as one of the examples of holistic assessments aimed at capturing the notion of 'co-evolution of dynamic partial pattern of unmistakable character' beyond the limits of a strictly regional, sectoral or process-oriented approach (Schellnhuber, Block, Cassel-Gintz, Kropp, Lammel, Lass, Lienenkamp, Loose, Lüdeke, Moldenhauer, Petschel-Held, Plöchl and Reusswig, 1997). Hereby, many factors are seen to flow together signifying sort of 'misdevelopments in the recent history of civilization-nature relations' or 'archetypical patterns of civilization-nature interactions', however, not to be interpreted as simple complexes of causes and effects. Nonetheless, a geographically explicit overview of GC could be obtained in the superposition of all mosaic structures showing the spatial distribution of single syndromes. Mapping done as such will not be a display of static patterns but more so the localization of active zones of problematic environmental and developmental processes. While in chapter 11 specific syndromes as well as methodological aspects of syndrome diagnosis will be explained in more detail, it is mentionend here that altogether 16 categories are further grouped by the type of human (mis)usage of natural resources, i.e., as a source of production ('utilization'), a medium for socio-economic development, or a sink for civilizational output.

Transects   The transect approach is originally an IGBP concept that attempts to link sites occupying different positions along some (strongly biophysically defined) gradient relevant to global change such as land use, rainfall or temperature. The emphasis of transects is on regional-scale impacts, while it has to be borne in mind that transects are conceptual. Thus, sites will not necessarily need to fall along a geographical line and a network of sites would work just as well.

The idea behind is simply the expectation that when analysed together, the range of sites is assumed to provide more and contingent information about the response of ecosystems to changes in the major gradients identified than if the sites were analysed alone. To be located along a transect are study sub-regions encompassing a range of characteristic features of the 'life-support system' (Odum, 1989), i.e., land-use practices as well as natural system properties. A study sub-region (<200 x 200 km) will contain at least one study site of 10 x 10 km in extent. The spatial dimensions are mainly determined by the requirements of the reliable use of remote sensing.

It has to be noted, however, that when adopting the transect framework of research organization it is not meant at all that the political ecology information to be collected will be covered by the prevailing Global Terrestrial Observing System (GTOS) protocol in its present state (Heal, Menaut and Steffen, 1993). Just the opposite holds true. A suitable platform to start with extending the hitherto mainly agronomical information collected by GTOS is seen in a major reworking of particularly the variable complex 1 ('site history and disturbance') by including, for example, the livelihood systems of rural land managers. With the newly emerging Land-Use and Land-Cover Change (LUCC) project being located at the interface of IGBP and IHDP (Turner, Skole, Sanderson, Fischer, Fresco and Leemans, 1995), the land manager approach is seen to be easily linked to the study of the driving forces of land-use and land-cover changes (Ehlers, 1998; Manshard, 1998 b).

Geography, which is positioned between social sciences on the one hand (human geography) and natural sciences on the other hand (physical geography), is seen to be particularly qualified to address GEC challenges outlined as such (Spada and Scheuermann, 1998, p. 84).

> Human geography, although traditionally focussing its empirical research on local and regional levels, increasingly also links its work to wider global contexts. So, on the one hand, human geography can take into account real people with their management and survival strategies and real environments with their problems of soil degradation, loss of biodiversity, deforestation, etc.,

but on the other hand, can also link these with the wider driving forces of global environmental change. From this perspective, it is clear that interdisciplinary approaches are called for, and that geography may play an integrative role in bridging research activities from both social and natural sciences.

Sustainability   In conceptualizing modes of (un)sustainable livelihood security for the majority of people living at present in the neighborhoods and villages of Africa, Asia and Latin America, a three dimensional concept has to be brought into application the valuation of which in quantifiable terms still is under work (see in particular chapter 2). Much of it will imply 'uneconomic growth' (Daly and Cobb, 1989) which as a term seems to be self-contradictory only from the perspective of mainstream neo-classical economics. Also, obvious policy implications will result in that sustainable development (beyond the blurred notion of WCED) means development without throughput growth beyond environmental carrying capacity and which is socially sustainable demanding increasing investments in human resources particulary in those of poor and marginalised people of the developing world (Manshard, 1993; Arts, 1994; Goodland, 1998). The special context and significance of the 'generational' aspects of sustainability within the 'Third-world' context was pointed out by Goodland (1998, p. 5).

> Most people in the world today are either impoverished or live barley above subsistence; the number of people living in poverty is increasing ... Future generations seem likely to be larger and poorer than today's generation. Even if the human population starts to decline after c.2050, it will inherit and have to make do with damaged life-support systems. How damaged is up to us of today's generation. Sustainability includes an element of not harming the future (intergenerational equity), as well as not harming society today (intra-generational equity). If the world cannot move toward intragenerational sustainability during this generation, it will be that much more difficult to achieve intergenerational sustainability sometime in the future. This is because the capacity of environmental services are being impaired, so will likely lower in the future than they are today.

In terms of definition, social sustainability will sharply be different from environmental and economic sustainability both having especially strong linkages. Also, the fundamental point about environmental sustainability is that it is a natural science concept and obeys biophysical laws making the definition robust irrespective of country, sector, or future epoch (Goodland and Daly, 1996).

Social sustainability is seen to be only achievable by systematic community participation and strong civil society. The many and hardly measurable components of 'social (or moral) capital' such as cohesion, identity and commonly accepted standards of honesty require maintenance and replenishment by shared values and equal rights, by community, religious and cultural interactions. If not cared for, social capital depreciates as surely as does physical capital. Though investment in social or human capital (e.g., education, health and nutrition) is now accepted as part of development, the creation and maintenance of it for social sustainability, however, is not (yet) acknowledged (Serageldin, Daly and Goodland, 1995; Goodland, 1998).

Economic sustainability relates to the 'maintenance of capital', i.e., keeping capital intact, but is rarely concerned with natural (resource) capital. To the criteria of 'allocation' and 'efficiency' a third is increasingly seen to be added which is 'scale', i.e., constraining throughput growth which is the flow of material and energy (natural capital) from environmental sources to sinks (Serageldin, Daly and Goodland, 1995; Goodland, 1998).

Environmental sustainability in the form of 'maintance of natural capital' could be expanded as input/output rules. Two fundamental environmental services have to be maintained unimpaired during the period over which sustainability is required, i.e., the use of (un)renewable resources on the source side and pollution and waste assimilation on the sink side (Daly and Cobb, 1989; Goodland, 1998). What is called the (normative) Serafian quasi-sustainability rule (or user cost approach) pertains to (non-)renewable resources to the extent they are being 'mined'. It states that resource owners may enjoy part of the proceeds from their liquidation as income (e.g., for consumption), while the remainder (a user cost) should be reinvested to produce income that would continue after the resource has been exhausted (i.e., no reinvestment in any asset that would produce future income, but specifically to produce renewable substitutes for the asset being depleted) (El Serafy, 1989).

Vulnerability   It has been mentioned earlier that among the key concepts and issues of ecological endangerment, an array of often overlapping or conflicting terms has arisen of which vulnerability is just one. The various aspects and theoretical concepts underlying the socially defined term and relating to GEC are seen to be well covered – among others – by Manshard (1991), Watts and Bohle (1993 a, b) and Bohle, Downing and Watts (1994). Most of the case studies explicitly exploring the dimensions of

vulnerability will relate to Chambers (1989, p. 1) who defines vulnerability
as

> exposure to contingencies and stress, and difficulties in coping with them.
> Vulnerability has thus two sides: an external side of risks, shocks, and stress to
> which an individual or household is subject; and an internal side which is
> defencelessness, meaning a lack of means to cope without damaging loss. Loss
> can take many forms – becoming or being physically weaker, economically
> impoverished, socially dependent, humiliated or psychologically harmed.

From the need for a patient and sensitive learning from those who are
vulnerable and poor together with a need for research directed towards
finding 'ways of strengthening and supporting people's present strategies for
coping' (Chambers, 1989), a growing body of case studies emerged that
approached the often and at first glance invisible strategies of vulnerable
people to cope with endangered livelihoods – and next to all of the chapters in
the volume contribute to this. In particular, the case studies dealing with
charcoal producers such as in Tanzania (chapter 6) and the Dominican
Republic (chapter 10) will directly respond, for example, to Manshard's
(1991, p. 286-7) admonition that the 'food-energy nexus' had not been on the
agenda of UNU framing the understanding of environmental degradation,
deforestation and vulnerability of rural and urban groups.

> Another type of vulnerable group, that has attracted the attention of many
> scholars of the rural poverty syndrome, is related to the low level of energy
> production. In most cases, for instance, charcoal producers belong to the poorest
> of the strata of rural society, often immigrant labour, who do not have many other
> options to make a living (...) They will cut down the trees when they feel the
> prices paid to them on the markets (mostly urban) are attractive enough to cover
> the cost and make a small profit. Both rural and urban charcoal consumers, on
> the other hand, use this energy source since it is the cheapest available alternative
> to which they are traditionally accustomed. Again it is rather the lower or middle
> level of society that are the users because electricity and gas are too expensive
> for them.

It is not misleading to state that a hitherto unmatched broad community
of social science researchers and policy makers got involved into the idea
that people's coping strategies do not only serve as useful indicators of crisis
(thus functioning as early warning systems), but that they also constitute
important means of reducing vulnerability and are henceforth worth to be
strengthened and supported. At this stage of conceptual exploration,
however, two points have to be raised starting from empirical evidence but

also emplying methodological (if not epistemological) consequences related to the social science concept as outlined.

Goodland (1998) notes on the concept of environmental sustainability that the growth of throughput, i.e., the flow of materials and energy from the sources of the environment, used by humans and then returned to environmental sinks as waste, translates into ever increasing rates of resource extraction and pollution or the use of sources and sinks. This is why the definition of unsustainability arises in that 'the scale of throughput has exceeded environmental source and sink capacities (and) the evidence is pervasive', namely, accumulation of green house gases, ozon shield depletion, pollution of drinking water and decline of natural forest cover. With special reference to the situation in the developing world, he further adds 'that the number of poor people is increasing suggests low cost extractive or harvested foods have become scarce or unobtainable'. In other words, this will directly affect the coping strategies of vulnerable people. While it is a methodologically tricky enterprise to combine the social science concept of vulnerability and livelihood systems with the natural science concept of environmental sustainability, it is nonetheless done in the following.

Chamber's (1989) social science definition of the internal side of vulnerability ('lack of means to cope without damaging loss') is linked to the natural science definition of environmental sustainability, namely, coping strategies that do not follow the Serafian rule of quasi-sustainability, i.e., 'mining' (non-)renewable resources without mitigation. The point is made that under the state of defencelessness the respective 'defence mechanisms' applied in natural resource management could turn to damaging losses both on the output side (e.g., waste emissions) as well as on the input side (e.g., harvest rates above the regenerative capacity of the natural system), while the (ecological) losses cannot be mitigated internally due to the many aspects of (social) losses already occured (under the general state of defencelessness). Thus, the possibility of long-term harmful effects of defence mechanisms against risks, shocks and stress upon the environment will be eminent, and so is the capacity of socially deprived people to detrimentally change and 'mine' their natural environment as a consequence of their vulnerability.

Environment and coping strategies The nature of coping strategies is mostly considered to be short-term and aimed at future entitlement security (Cannon, 1991, p. 304).

> The term coping strategies ... relates mainly to the activities of people involved
> directly in production of much of their own needs, and implies forms of

behaviour which are abnormal. They are in effect a set of reserve or emergency entitlements. Coping involves not only dealing with ... threat, but also may be aimed at the preservation of people's livelihood systems (including their entitlements).

The use of (short-term) coping strategies – as contrasted with survival strategies – may become long-term where conditions of production and exchange have shifted permanently to the detriment of a group of people. This is to explore further the point raised by Cannon (1991) that vulnerable people might deal with disruptions as such 'by various coping strategies, sometimes with damaging impact on the environment'.

Against the background of structural violence (Galtung, 1969, 1975, 1996), i.e., predispositioning or 'systemic' disparate social structures marginalising an assumedly large and increasing number of people and preventing them from the effective implementation of life chances, a crisis of livelihood systems could be seen as the culminating point of long-term processes that have to be taken under consideration. The notion was exemplified by Watts (1989, 1990) who described the African agrarian crisis as the result of long-term processes embedded in disparities and conflicts which are the outcome of the contradictory and 'critical' (i.e., spatially and socially polarising) development of the global economic system. The process could be taken as paradigmatic and seen to lead to structural weakness, instability and crisis proneness of livelihood systems in large (if not most) parts of the developing world.

At this point of exploration, and drawing much on the findings of chapter 4, the notion of environmentally harmful coping strategies has to be raised. Though coping by definition implies sort of effectively managing a situation, the term 'harmful coping' is seen to be (merely) self-contradictory as it is 'uneconomic growth'. Inherent in the normally short-term intended coping mechanisms could be measures that severely deplete those environmental resources which are crucial to the livelihood of affected people. Cannon (1991, p. 305-6) pointed out that

> such coping is designed to preserve livelihood assets (...) and may occur such that the mechanism used by one person (or) household would be well designed for their maintenance were it not for the fact that many others have to pursue the same strategy at the same time. In such a situation, a very reasonable and rational survival mechanism for the private preservation of the individual becomes a seemingly irrational and self-destructive form of behaviour for the wider community.

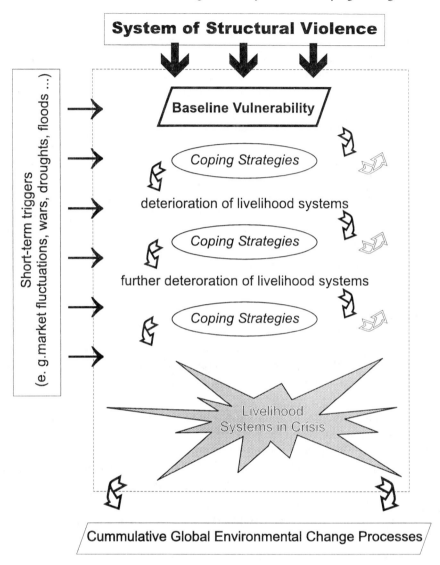

Effects of environmentally neutral or positive strategies

Effects of environmentally harmful strategies

**Figure 1.2   Framework of structural violence and harmful coping
strategies**

In the form of Blaikie's (1985) work on the political economy of soil erosion, this became integral part of policial ecology modelling, i.e., coping mechanisms really are no longer short-term measures, but become long-term and normal methods of existence in that 'people are forced to destroy their own and their decendants' future in order to survive' (Cannon, 1991).

The framework of understanding the interaction between what is called here structural violence and coping is given in Figure 1.2. It implies that grassroots actors or people at the local level react on vulnerability enforcing processes by applying coping strategies which set in motion further deterioration of the natural environment and, by 'mining' it, even enhance their vulnerability further. Using this notion it is borne in mind that 'difference' allows for winners and loosers and, thus, the political-ecological paradox is broached not only why environmental change poses a problem at all, but also for whom.

Conclusions

By outlining the political ecology framework as a most suitable research perspective on the human dimensions of global environmental change (GEC), it is not meant at all to rank lower any other attempts such as 'landscape ecology', 'metabolism', 'eco-restructuring' or 'earth systems analysis' but more so to point to the fact that any serious discussion of social relations involved has to start with a notion of 'structure' or 'difference'. What has hitherto been conceptualized in terms of 'institutional' (say, social) dimensions of GEC is seen not to reflect at all the livelihood systems and endangerments of the vast majority of people living in the neighborhoods and villages of Africa, Asia and Latin America. And the worlds of neighborhoods and villages in the 'south' can not be separated from the 'world of comfort' in the 'north' as they are sometimes conveniently portrayed.

Following the lines of structural violence, and taking into account the negative aspects of coping with environmental change, the theorizing of vulnerability, coping and environmental change might be fruitfully complicated by the outline as given here. Relating to GEC as it occurs in the 'Third-world' context, at least two questions of endangerment should be added on the agenda. First, what are the coping strategies that turn out to be harmful for individuals and society as a whole, i.e., enhancing vulnerability in the long-run instead of reducing it? Secondly, what is the specific nature of the underlying causes for the implementation of environmentally harmful strategies? Though difficult in nature, any quantification of the issues outlined will be essential. This is seen on the background that it was early recognized by a study group of the International Geographical Union (IGU) on Critical Zones in Global Environmental Change (Kasperson, 1992, p. 31) that

> a critical environmental zone is much more likely to develop in poor or underdeveloped economies where the opportunities for economic and technological substitution are few, where global economic relations cause 'net' environmental degradation, and where the societal capacity to respond to the threat is limited.

Both in terms of conceiving scales of social reserach on GEC (functional, conceptual, spatial and temporal) and in terms of bridging 'misfits' between institutions and ecosystems, there is a need for porperly addressing the issues raised and faced by the majority of people living in 'Third-world' situations.

Grounding ecology in the web of social relations and following the demand for careful local-level studies, the chapters in the volume are seen to contribute adequately to this. The case studies and the concluding chapter, in their many different ways, provide perspectives on this question. They will provide sound material and some answers not always consistently to the questions and framework we have raised in this opening chapter, but they will also take the arguments much further.

Notes

[1]   The commission's report is seen to have replaced (and hereby forgotten about) the stronger economically, socio-politically and 'south' oriented view of the North-South Commission under Willy Brandt (Independent Commission on International Development Issues, 1980).

[2]   A much broader view underlies the concept of political ecology as outlined by Keil, Bell, Penz and Fawcett (1998, p. 13). They note that the editors of the inaugural volume of 'The Journal of Political Ecology' define political ecology as 'a historical outgrowth of the central questions asked by the social sciences about the relations between human society, viewed in its bio-cultural-political complexity, and a significantly human nature'. The two strands flowing together in political ecology are specified as 'political economy with its insistence on the need to link the distribution of power with productive activity and ecological analysis with its broader vision of bio-environmental relationships'; Greenberg, J.B. and Park, T.K. (1994), The Journal of Political Ecology, vol. 1, p. 1.

3   While this chapter was compiled and edited by H. Spada and M. Scheuermann, individual contributions were done by H. Birg, H.-G. Bohle, M. Casimir, H. Elsenhans, A. Endres, D. Frey, H. Geist, R. Guski, R. Höger, E. Matthies, E. Mohr, A. Obser, P. Pansegrau, K. Pawlik, J. Rost, M. Scheuermann, H. Spada, R. Wahl, P. Weingart and H. Zilleßen.

4   Citation follows draft versions of 1994 (Rauch) and 1997 (Goodland).

References

Altvater, E. (1998), 'Global Order and Nature', in R. Keil, D.V.J. Bell, P. Penz, P. and L. Fawcett (eds), *Political Ecology: Global and Local*, Routledge, London, New York, pp. 19-45.

Ante, U. (1985), *Zur Grundlegung des Gegenstandsbereiches der politischen Geographie: Über das 'Politische' in der Geographie*, Erdkundliches Wissen No. 75, Steiner, Wiesbaden.

Arts, B. (1994), 'Nachhaltige Entwicklung: Eine begriffliche Abgrenzung', *Peripherie*, vol. 14 (54), pp. 6-27.

Ayres, R.U. and Simonis, U.E. (eds) (1994), *Industrial Metabolism: Restructuring for Sustainable Development*, United Nations University Press, Tokyo, New York, Paris.

Bahro, R. (1989), *Logik der Rettung: Wer kann die Apokalypse aufhalten? Ein Versuch über die Grundlagen ökologischer Politik*, Weilbrecht, Stuttgart, Wien.

Bassett, T.J. (1988), 'The Political Ecology of Peasant-Herder Conflicts in the Northern Ivory Coast', *Annals of the Association of American Geographers*, vol. 78 (3), pp. 453-72.

Biot, Y., Blaikie, P., Jackson, C. and Palmer-Jones, R. (1995), *Rethinking Research on Land Degradation in Developing Countries*, World Bank Discussion Paper No. 289, World Bank, Washington, DC.

Blaikie, P. (1985), *The Political Economy of Soil Erosion in Developing Countries*, Longman, New York.

Blaikie, P. (1986), 'Natural Resource Use in Developing Countries', in R.J. Johnston and P.J. Taylor (eds), *A World in Crisis? Geographical Perspectives*, Basil Blackwell, Oxford, New York, pp. 107-26.

Blaikie, P. (1988), 'Environmental Crisis in Developing Countries: How much, for whom and by whom? An Introduction and Overview', in P. Blaikie and T. Unwin (eds), *Environmental Crises in Developing Countries*, DARG Monograph No. 5, Institute of British Geographers, London, pp. 1-6.

Blaikie, P. (1994), *Political Ecology in the 1990s: An Evolving View of Nature and Society*, CASID Distinguished Speaker Series No. 13, Michigan State University, Center for Advanced Study of International Development, East Lansing.

Blaikie, P. (1995), 'Changing Environments or Changing Views? A Political Ecology for Developing Countries', *Geography*, vol. 80 (3), pp. 203-14.

Blaikie, P. and Brookfield, H. (1987), *Land Degradation and Society*, Routledge, London, New York.

Blaikie, P., Cannon, T., Davis, I. and B. Wisner (1994), *At Risk: Natural Hazards, People's Vulnerability, and Disasters*, Routledge, London, New York.

Bohle, H.-G., Downing, T.E. and Watts, M.J. (1994), 'Climate Change and Social Vulnerability: Toward a Sociology and Geography of Food Insecurity', *Global Environmental Change*, vol. 4 (1), pp. 37-48.

Bryant, R.L. (1992), 'Political Ecology: A Emerging Research Agenda in Third-World Studies', *Political Geography*, vol. 11 (1), pp. 12-36.

Bryant, R.L. (1997), 'Beyond the Impasse: The Power of Political Ecology in Third World Environmental Research', *Area*, vol. 29 (1), pp. 5-19.

Bryant, R.L. and Bailey, S. (1997), *Third World Political Ecology*, Routledge, London, New York.

Bundesministerium für Umwelt, Naturschutz und Reaktorsicherheit (1992), *Umweltpolitik, Konferenz der Vereinten Nationen für Umwelt und Entwicklung im Juni 1992 in Rio de Janeiro, Dokumente: Agenda 21*, BMU, Bonn.

Cannon, T. (1991), 'Hunger and Famine: Using a Food System's Model to Analyse Vulnerability', in H.-G. Bohle, T. Cannon, G. Hugo and F.N. Ibrahim (eds), *Famine and Food Security in Africa and Asia: Indigenous Response and External Intervention to avoid Hunger*, Bayreuther Geowissenschaftliche Arbeiten, No. 15, Naturwissenschaftliche Gesellschaft, Bayreuth, pp. 291-312.

Chambers, R. (1989), 'Editorial Introduction: Vulnerability, Coping and Policy', *Institute of Development Studies Bulletin*, vol. 20 (2), pp. 1-7.

Daly, H.E. and Cobb, J. (1989), *For the Common Good: Redirecting the Economy toward Community, the Environment and a Sustainable Future*, Beacon Press, Boston.

Ehlers, E. (1998), 'Global Change und Geographie', *Geographische Rundschau*, vol. 50 (5), pp. 273-76.

Ehlers, E. and T. Krafft (1998), 'German Global Change Research: The Need for an Integrative Approach', in E. Ehlers and T. Krafft (eds), *German Global Change Research 1998*, German National Committee on Global Change Research, Bonn, pp. 6-8.

El Serafy, S. (1989), 'The Proper Calculation of Income from Depletable Natural Resources', in Y. Ahmad (ed), *Environmental Accounting for Sustainable Development*, World Bank, Wahsington, DC, pp. 10-8.

Emel, J. and Peet, R. (1989), 'Resource Management and Natural Hazards', in R. Peet and N. Thrift (eds), *New Models in Geography: The Political-Economy Perspective*, Unwin Hyman, London, Boston, Sydney, Wellington, pp. 49-76.

Enzensberger, H.M. (1973), 'Zur Kritik der politischen Ökologie', *Kursbuch*, vol. 33, pp. 1-42.

Enzensberger, H.M. (1974), 'A Critique of Political Ecology', *New Left Review*, vol. 8 (4), pp. 3-32.

Ernst, A.M. (1998), 'Umweltwandel und Allmende-Problematik: Ein Konzept leitet interdisziplinäre Forschung', *GAIA: Ecological Perspectives in Science, Humanities and Economics*, vol. 7 (4), pp. 251-54.

Fischer-Kowalski, M., Haberl, H., Hüttler, W., Payer, H., Schandl, H., Winiwarter, V. and Zangerl-Weisz, H. (eds) (1997), *Gesellschaftlicher Stoffwechsel und Kolonisierung von Natur: Ein Versuch in Sozialer Ökologie*, G+B Verlag Fakultas & Overseas Publishers Association, Amsterdam.

FitzSimmons, M., Glaser, J., Monte Mor, R., Pincetl, S. and Rajan, S.C. (1994), 'Environmentalism and the Liberal State', in M. O'Connor (ed), *Is Capitalism Sustainable? Political Economy and the Politics of Ecology*, Guilford Press, New York, London, pp. 198-216.

Franke, R.W. and Chasin, B.H. (1980), *Seeds of Famine: Ecological Destruction and the Development Dilemma in the West African Sahel*, Allenheld, Osmun, Mountclair.

48   Coping with Changing Environments

Friedrich, E. (1904), 'Wesen und geographische Verbreitung der "Raubwirtschaft"', *Petermanns Geographische Mitteilungen*, vol. 50, pp. 68-79, 92-5.

Galtung, J. (1969), 'Violence, Peace and Peace Research', *Journal of Peace Research*, pp. 167-91.

Galtung, J. (1975), *Strukturelle Gewalt: Beiträge zur Friedens- und Konfliktforschung*, Rowohlt, Reinbek bei Hamburg.

Galtung, J. (1996), *Peace by Peaceful Means: Peace and Conflict, Development and Civilisation*, PRIO & Sage, London, Thousand Oaks, New Delhi.

Geist, H. (1992), 'Die orthodoxe und politisch-ökologische Sichtweise von Umweltdegradierung', *Die Erde*, vol. 123 (4), pp. 283-95.

Geist, H. (1996), 'Political Ecology of the Lower Casamance in Senegal (West Africa): Debating the crisis', in R.B. Singh (ed), *Disasters, Environment and Development*, Oxford & IBH Publishers, New Delhi, Calcutta, pp. 541-50.

Geist, H. (1998), 'Das Bergland von Namwera: Eine Fallstudie über Landdegradierung, Gemeinheitsteilung und braunes Gold', *GAIA: Ecological Perspectives in Science, Humanities and Economics*, vol. 7 (4), pp. 255-64.

Gibson, C., Ostrom, E. and Ahn, T.-K. (1998), *Scaling Issues in the Social Sciences*, IHDP Working Paper No. 1, International Human Dimensions Programme on Global Environmental Change, Bonn.

Giddens, A. (1984), *The Constitution of Society: Outline of the Theory of Structuration*, Polity Press, Cambridge.

Giddens, A. (1993), *Sociology*, Polity Press, Cambridge.

Goodland, R. (1998), 'The Biophysical Basis of Environmental Sustainability', in J.C.J.M. van den Bergh (ed), *Handbook of Environmental and Resource Economics*, Edward Elgar, London.

Goodland, R. and Daly, H.E. (1996), 'Environmental Sustainability: Universal and Non-negotiable', *Ecological Applic*, vol. 6, pp. 1002-1017.

Grossman, L.S. (1998), *The Political Ecology of Bananas: Contract Farming, Peasants and Agrarian Change in the Eastern Caribbean*, University of North Carolina Press, Chapel Hill, London.

Hambloch, H. (1986), *Der Mensch als Störfaktor im Geosystem*, Vorträge der Rheinisch-Westfälischen Akademie der Wissenschaften/Geisteswissenschaften No. 280, Westdeutscher Verlag, Opladen.

Hannah, L., Lohse, D., Hutchinson, C., Carr, J.L. and Lankerani, A. (1994), 'A Preliminary Inventory of Human Disturbance of World Ecosystems', *Ambio*, vol. 23 (4/5), pp. 246-50.

Hard, G. (1997), 'Was ist Stadtökologie? Argumente für eine Erweiterung des Aufmerksamkeitshorizonts ökologischer Forschung', *Erdkunde*, vol. 51, pp. 100-13.

Hardesty, D. (1977), *Ecological Anthropology*, John Wiley, New York.

Hardin, G. (1968), 'The Tragedy of the Commons', *Science*, vol. 162, pp. 1243-8.

Hardin, G. (1970), 'Die Tragik der Allmende', in M. Lohmann (ed), *Gefährdete Zukunft: Prognosen angloamerikanischer Wissenschaftler*, Carl Hanser, München, p. 30-48.

Hardin, G. (1993), *Living with Limits: Ecology, Economics and Population Taboos*, Oxford University Press, New York.

Harvey, D.W. (1969), *Explanation in Geography*, St. Martin's Press, New York.

Harvey, D.W. (1974), 'Population, Resources and the Ideology of Science', *Economic Geography*, vol. 50, pp. 256-77.

Harvey, D.W. (1996), *Justice, Nature and the Geography of Differences*, Basil Blackwell, Oxford.

Heal, O.W., Menaut, J.-C. and Steffen, W.L. (1993), *Towards a Global Terrestrial Observing System (GTOS): Detecting and Monitoring Change in Terrestrial Ecosystems*, MAB Digest No. 14, IGBP Global Change Report, No. 26, UNESCO, IGBP, Paris, Stockholm.

Heberle, R. (1945), *From Democracy to Nazism*, Baton Rouge.

Heberle, R. (1978), 'Wahlökologie', in R. König (ed), *Handbuch der empirischen Sozialforschung*, No. 12, Enke, Stuttgart, pp. 73-102.

Hecht, S.B. (1985), 'Environment, Development and Politics: Capital Accumulation and the Livestock Sector in Eastern Amazonia', *World Development*, vol. 13, pp. 663-84.

Independent Commission on International Development Issues (1980), *North-South: A Programme for Survival*, Kiepenheuer & Witsch, Köln.

Intergovernmental Panel on Climate Change (1990), *Climate Change: The IPCC Scientific Assessment*, Cambridge University Press, Cambridge.

Johnston, R.J. (1986), *On Human Geography*, Blackwell, Oxford.

Johnston, R.J., Taylor, P.J. and O'Loughlin, J. (1987), 'The Geography of Violence and Premature Death: A World Systems Approach', in R. Vayrynen (ed), *The Quest for Peace*, Sage, London, pp. 241-59.

Johnston, R.J., Taylor, P.J. and Watts, M.J. (1995 a), 'Introduction to Part V (Geoenvironmental Change): A Burden too Far?', in R.J. Johnston, P.J. Taylor and M.J. Watts (eds), *Geographies of Global Change: Remapping the World in the late Twentieth Century*, Blackwell, Oxford, pp. 297-301.

Johnston, R.J., Taylor, P.J. and Watts, M.J. (1995 b), 'Introduction to Part III (Geosocial Change): People in Turmoil', in R.J. Johnston, P.J. Taylor and M.J. Watts (eds), *Geographies of Global Change: Remapping the World in the late Twentieth Century*, Blackwell, Oxford, pp. 147-51.

Kasperson, R.E. (1992), 'Human Response to Environmental Degradation in En-dangered Areas', *Acta Universitatis Carolinae Geographica*, vol. 1, pp. 29-36.

Kasperson, R.E. (1993), 'Critical Environmental Regions and the Dynamics of Change', in H.-G. Bohle, T.E. Downing, J.O. Field and F.N. Ibrahim (eds), *Coping with Vulnerability and Criticality: Case studies on Food-Insecure People and Places*, Freiburg Studies in Development Geography No. 1, Breitenbach, Saarbrücken, Ft. Lauderdale, pp. 115-26.

Kasperson, J.X., Kasperson, R.E. and Turner, B.L.II (eds), *Regions at Risk: Comparisons of Threatened Environments*, United Nations University Press, Tokyo, New York, Paris.

Kasperson, R.E., Kasperson, J.X, Turner, B.L.II, Dow, K. and Meyer, W.B. (1995), 'Critical Environmental Regions: Concepts, Distinctions, and Issues', in J.X. Kasperson, R.E. Kapserson and B.L. Turner II (eds), *Regions at Risk: Comparisons of Threatened Environments*, United Nations University Press, Tokyo, New York, Paris, pp. 1-41.

Kautsky, K. (1927), *Die materialistische Geschichtsauffassung: Natur und Gesellschaft*, No. 1, Dietz Nachfolger, Berlin.

Kautsky, K. (1899/1966), *Die Agrarfrage: Eine Übersicht über die Tendenzen der modernen Landwirtschaft und die Agrarpolitik der Sozialdemokratie*, Sozialistische Klassiker in Neudrucken No. 9, Dietz Nachfolger, Hannover.

Kautsky, K. (1988), *The Agrarian Question*, Zwan, London.

Keil, R., Bell, D.V.J., Penz, P. and Fawcett, L. (1998), 'Editors' Introduction: Perspectives on Global Political Ecology', in Keil, R., Bell, D.V.J., Penz, P. and Fawcett, L. (eds), *Political Ecology: Global and Local*, Routledge, London, New York, pp. 1-16.

Krings, T. (1994), 'Theoretische Ansätze zur Erklärung der ökologischen Krise in der Sahelzone', *Zeitschrift für Wirtschaftsgeographie*, vol. 38 (1/2), pp. 1-10.

Krings, T. (1996), 'Politische Ökologie der Tropenwaldzerstörung in Laos', *Petermanns Geographische Mitteilungen*, vol. 140 (3), pp. 161-75.

Little, P.D. and Watts, M.J. (eds), *Living under Contract: Contract Farming and Agrarian Transformation in Sub-Saharan Africa*, University of Wisconsin Press, Madison.

Lohnert, B. (1995), *Überleben am Rande der Stadt: Ernährungssicherungspolitik, Getreidehandel und verwundbare Gruppen in Mali, Das Beispiel Mopti*, Freiburg Studies in Development Geography No. 8, Verlag für Entwicklungspolitik, Saarbrücken.

Lohnert, B. (1998), 'Die Politische Ökologie der Land-Stadt Migration in Südafrika', *GAIA: Ecological Perspectives in Science, Humanities and Economics*, vol. 7 (4), pp. 265-70.

Manshard, W. (1991), 'Sustainable Development, Global Programmes, and Famine Research', in H.-G. Bohle, T. Cannon, G. Hugo and F.N. Ibrahim (eds), *Famine and Food Security in Africa and Asia: Indigenous Response and External Intervention to avoid Hunger*, Bayreuther Geowissenschaftliche Arbeiten No. 15, pp. 279-90.

Manshard, W. (1993), 'Toward Sustainable Development: Progress of the International Global Change Programmes', in H.-G. Bohle, T.E. Downing, J.O. Field and F.N. Ibrahim (eds), *Coping with Vulnerability and Criticality: Case studies on Food-Insecure People and Places*, Freiburg Studies in Development Geography No. 1, Breitenbach, Saarbrücken, Ft. Lauderdale, pp. 361-76.

Manshard, W. (1998 a), 'The Biophysical Basis of Eco-Restructuring: An Overview of Current Relations between Human Economic Activities and the Global System', in R.U. Ayres and P.M. Weaver (eds), *Eco-Restructuring: Implications for Sustainable Development*, United Nations University Press, Tokyo, New York, Paris, pp. 55-76.

Manshard, W. (1998 b), 'Bevölkerung, Landnutzung und Umweltwandel in den Tropen', *Geographische Rundschau*, vol. 50 (5), pp. 278-82.

Manshard, W. and Mäckel, R. (1995), *Umwelt und Entwicklung in den Tropen: Naturpotential und Landnutzung*, Wissenschaftliche Buchgesellschaft, Darmstadt.

McDonald, M.D. (1993), 'Dams, Displacement, and Development: A Resistance Movement in Southern Brazil', in J. Friedmann and H. Rangan (eds), *In Defense of Livelihood: Comparative Studies on Environmental Action*, Kumarian Press, West Hartford.

Meentemeyer, V. (1989), 'Geographical Perspectives of Space, Time, and Scale', *Landscape Ecology*, vol. 3 (3/4), pp. 163-73.

Meyer, W.B., Derek, G., Turner, B.L.II and McDowell, P.F. (1992), 'The Local-Global Continuum', in: R.F. Abler, M.G. Marcus and J.M. Olson (eds), *In Geography's Inner Worlds*, Rutgers University Press, New Brunswick, pp. 255-79.

Meyer, W.B. and Turner, B.L.II (1992), 'Human Population Growth and Global Land-use/Land-cover Change', *Annual Review of Ecology and Systematics*, vol. 23, pp. 39-61.

Meyer, W.B. and Turner, B.L. II (1995), 'The Earth Transformed: Trends, Trajectories and Patterns', in R.J. Johnston, P.J. Taylor and M.J. Watts (eds), *Geographies of Global Change: Remapping the World in the Late Twentieth Century*, Blackwell, Oxford, Malden, pp. 302-17.

Mitchell, J.K. (1998), 'Urban Metabolism and Disaster Vulnerability in an Era', in H.-J. Schellnhuber and V. Wenzel (eds), *Earth System Analysis: Integrating Science for Sustainability*, Springer, Berlin, Heidelberg, New York, pp. 359-77.

Moore, B.III (1996), 'Global Models: Sooner rather than Later', *IGBP Newsletter*, vol. 26, pp. 11-2.

Mumford, L. (1977), *Mythos der Maschine: Kultur, Technik und Macht, Die umfassende Darstellung der Entdeckung und Entwicklung der Technik*, Fischer, Frankfurt/M.

Netting, R. (1986), *Cultural Ecology*, Waveland Press, Prospect Heights.

Noble, I.R. (1996), 'Linking the Human Dimension to Landscape Ecology', in B. Walker and W. Steffen (eds), *Global Change and Terrestrial Ecosystems*, IGBP Book Series, Cambridge University Press, Cambridge, pp. 173-83.

O'Connor, M. (ed) (1994), *Is Capitalism Sustainable? Political Economy and the Politics of Ecology*, Guilford Press, New York, London.

Odum, E.P. (1989), *Ecology and our Endangered Life-Support Systems*, Sinauer Associates, Sunderland.

Peet, R. and Thrift, N. (1989), 'Political Economy and Human Geography', in R. Peet and N. Thrift (eds), *New Models in Geography, Volume 1: The Political-Economy Perspective*, Unwin Hyman, London, pp. 3-29.

Pritchard, L., Colding, J., Berkes, F., Svedin, U. and C. Folke (1998), *The Problem of Fit between Ecosystems and Institutions*, IHDP Working Paper No. 2, International Human Dimensions Programme on Global Environmental Change, Bonn.

Quarry, J. (ed) (1992), *Earth Summit*, Regency Press, London.

Ramankutty, N. and J. Foley (1998), 'Characterizing Patterns of Global Land Use: An Analysis of Global Cropland Data', *Global Biogeochemical Cycles*, vol. 12, pp. 667-85.

Ramankutty, N. and J. Foley (1999), *Estimating Historical Changes in Global Land Cover: Croplands from 1700 to 1992*, University of Wisconsin, Madison.

Rauch, T. (1996), *Ländliche Regionalentwicklung im Spannungsfeld zwischen Weltmarkt, Staatsmacht und kleinbäuerlichen Strategien*, Sozialwissenschaftliche Studien zu internationalen Problemen No. 202, Verlag für Entwicklungspolitik, Saarbrücken.

Rauch, T., Haas, A. and Lohnert, B. (1996), 'Ernährungssicherheit in ländlichen Regionen des tropischen Afrika zwischen Weltmarkt, nationaler Agrarpolitik und den Sicherungsstrategien der Landbevölkerung', *Peripherie*, vol. 63, pp. 33-72.

Redclift, M. and T. Benton (eds) (1994), *Social Theory and the Global Environment*, Routledge, London, New York.

Schellnhuber, H.-J., Block, A., Cassel-Gintz, M., Kropp, J., Lammel, G., Lass, W., Lienenkamp, R., Loose, C., Lüdeke, M.K.B., Moldenhauer, O., Petschel-Held, G., Plöchl, M. and Reusswig, F. (1997), 'The Syndromes of Global Change', *GAIA: Ecological Perspectives in Science, Humanitites and Economics*, vol. 6 (1), pp. 19-34.

Schellnhuber, H.-J. and Wenzel, V. (1998), *Earth System Analysis: Integrating Science for Sustainability*, Springer, Berlin, Heidelberg, New York.

Schneider, H. and Geist, H. (1996), 'Erosions- und Reformprozesse in städtischen und ländlichen Räumen Afrikas: Beiträge geographischer Entwicklungsforschung', in P. Meyns (ed), *Staat und Gesellschaft in Afrika: Erosions- und Reformprozesse*, Schriften der Vereinigung von Afrikanisten in Deutschland No. 16, Lit, Hamburg, pp. 472-85.

Seers, D. (1974), 'Was heißt "Entwicklung"?', in D. Senghaas (ed), *Peripherer Kapitalismus: Analysen über Abhängigkeit und Unterentwicklung*, Suhrkamp, Frankfurt/Main, pp. 39-67.

Serageldin, I., Daly, H.E. and Goodland, R. (1995), 'The Concept of Sustainability', in W. van Dieren (ed), *Taking Nature into Account*, Springer, New York, pp. 99-123.

Smith, N. (1984), *Uneven Development: Nature, Capital and the Production of Space*, Basil Blackwell, Oxford, Cambridge.

Spada, H. and Scheuermann, M. (1998), 'The Human Dimensions of Global Environmental Change, Social Science Research in Germany', in E. Ehlers and T. Krafft (eds), *German Global Change Research 1998*, German National Committee on Global Change Research, Bonn, pp. 71-91.

Stern, P. C., Young, O.R., and Druckman, D. (eds) (1992), *Global Environmental Change: Understanding the Human Dimensions*, National Academy Press, Washington, DC.

Stern, P. C., Young, O.R., and Druckman, D. (eds) (1994), *Science Priorities for the Human Dimensions of Global Change*, National Academy Press, Washington, DC.

Taylor, P.J. (1981), 'Geographical Scales within the World Economy Approach', *Geographical Review*, vol. 3 (1), pp. 3-11.

Taylor, P.J., Watts, M.J. and Johnston, R.J. (1995), 'Global Change at the End of the Twentieth Century', in R.J. Johnston, P.J. Taylor and M.J. Watts (eds), *Geographies of Global Change: Remapping the World in the late Twentieth Century*, Blackwell, Cambridge, pp. 1-10.

Turner, B.L.II and Butzer, K.W. (1992), 'The Columbian Encounter and Land-use Change', *Environment*, vol. 43, pp. 16-20.

Turner, B.L.II, Clark, W.C., Kates, R.W., Richards, J.F., Mathews, J.T. and Meyer, W.B. (eds) (1990), *The Earth as Transformed by Human Action: Global and Regional Changes in the Biosphere over the past 300 Years*, Cambridge University Press, Clark University, Cambridge, New York, Port Chester, Melbourne, Sydney.

Turner, B.L.II, Kasperson, J.X., Kasperson, R.E., Dow, K. and Meyer, W.B. (1995), 'Comparisons and Conclusions', in J.X. Kasperson, R.E. Kasperson and B.L. Turner II (eds), *Regions at Risk: Comparisons of Threatened Environments*, United Nations University Press, Tokyo, New York, Paris, pp. 519-86.

Turner, B.L. II, Skole, D., Sanderson, S., Fischer, G., Fresco, L. and Leemans, R. (1995), *Land-Use and Land-Cover Change: Science/Research Plan*, IGBP Report No. 35, HDP Report No. 7), IGBP, HDP, Stockholm, Geneva.

United Nations Conference on Environment and Development (1992): *Agenda 21*, UNCED, Conches.

United Nations Environment Programme (1992), *Convention on Biological Diversity*, UNEP, Environmental Law and Institutions Programme Centre, Nairobi.

Walker, B. and Steffen, W. (eds) (1996), *Global Change and Terrestrial Ecosystems*, IGBP Book Series, Cambridge University Press.

Watts, M.J. (1984), 'The Demise of the Moral Economy: Food and Famine in a Sudano-Sahelian Region in a Historical Perspective', in E. Scott (ed), *Life before the Drought*, Allen & Unwin, London, pp. 124-48.

Watts, M.J. (1985), 'Social Theory and Environmental Degradation', in Y. Gradus (ed), *Desert Development: Man and Technology in Sparse Lands*, Reidel, Dordrecht, pp. 14-32.

Watts, M.J. (1989), 'The Agrarian Question in Africa: Debating the Crisis', *Progress in Human Geography*, vol. 13 (1), pp. 1-41.

Watts, M.J. (1990), 'Die Agrarfrage in Afrika: Debatten über die Krise', *Journal für Entwicklungspolitik*, vol. 1, pp. 5-57.

Watts, M.J. (1996), 'Development III: The Global Agrofood System and late twentieth-century Development (or Kautsky Redux)', *Progress in Human Geography*, vol. 20 (2), pp. 230-45.

Watts, M.J. and Bohle, H.G. (1993 a), 'The Space of Vulnerability: The Causal Structure of Hunger and Famine', *Progress in Human Geography*, vol. 17 (1), pp. 43-67.

Watts, M.J. and Bohle, H.-G. (1993 b), 'Hunger, Famine and the Space of Vulnerability', *GeoJournal*, vol. 30 (2), pp. 117-25.

Williams, M. (1994 a), 'Forests and Tree cover', in W.B. Meyer and B.L. Turner II (eds), *Changes in Land Use and Land Cover: A Global Perspective*, Cambridge University Press, Cambridge, pp. 97-124.

Williams, M. (1994 b), 'The Relations of Environmental History and Historical Geography', *Journal of Historical Geography*, vol. 20 (1), pp. 3-21.

Wolf, E. (1972), 'Ownership and Political Ecology', *Anthropological Quarterly*, vol. 45, pp. 201-5.

World Commission on Environment and Development (1987), *Our Common Future, The Brundtland Commission*, Oxford University Press, Oxford, New York.

Zimmerer, K.S. (1996), 'Ecology as Cornerstone and Chimera in Human Geography', in C. Earle, K. Mathewson and M.S. Kenzer (eds), *Concepts in Human Geography*, Rowman & Littlefield, Lanham, pp. 161-88.

Zimmermann, E.W. (1951), *World Resources and Industries*, Harper, New York.

2 Coping with Vulnerability and Unsustainability
The case of Nepalese upland farmers

HANS-GEORG BOHLE

Introduction

In a remarkable World Bank publication, Serageldin (1996) developed a new index to calculate the 'wealth of nations' and to link the calculation to global issues of environmental sustainability. The new indicators of sustainable development start from a definition of sustainability as 'opportunity' in that 'sustainability is to leave future generations as many opportunities as we ourselves have had, if not more'.

As a first step towards conceptualization, the idea of environmentally sustainable development is defined in terms of a triangular framework (Serageldin and Steer, 1994). As given in Figure 2.1, the three sides of the triangle represent the economic, ecological and social dimensions of sustainability. It is noted here by Serageldin (1996, p. 3) that

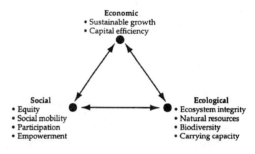

Figure 2.1   ESD Triangle

Source: Serageldin, I. and Steer, A. (eds) (1994), *Making Development Sustainable: From Concepts to Action*, World Bank, Washington DC, p. 2

a proposal has to be economically and financially sustainable in terms of growth, capital maintenance, and efficient use of resources and investments. But it also has to be ecologically sustainable, and here we mean ecosystem integrity, carrying capacity, and conservation of natural resources, including biodiversity. Ecological sustainability is the domain of the biologist and the physical scientist. The units of measurement are different, the constructs are different, and the context and time scale are different. However, equally important is the social side, and here we mean equity, social mobility, social cohesion, participation, empowerment, cultural identity, and institutional development. The social dimension is the domain of the sociologist, the anthropologist, and the political scientist. It is, to my mind, an essential part of the definition of sustainability, because the neglect of the social dimension leads to institutions that are incapable of responding to the needs of society.

From the perspective adopted by the World Bank, sustainability as opportunity translates into providing future generations with as much capital per capita as we ourselves have had. According to Figure 2.2, four kinds of capital could be distinguished, i.e., man-made capital, natural capital, human capital and social capital. While man-made capital is the one usually considered in economic accounts (e.g., built infrastructure), natural capital is basically the natural endowment being defined as the stock of environmentally provided assets (such as soil, atmosphere, forests, water and wetlands). Natural capital is then distinguished from other forms of capital, namely human capital (i.e., people, their education, their health and capacity levels). This form of capital is currently regarded to be of increasing importance since investment in people is now seen to constitute a very high-return investment particularly in developing countries. Finally, social capital recognizes the importance of social cohesion and of a functioning social order based on myriads of social institutions.

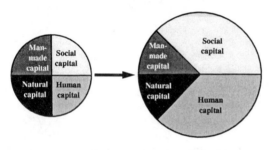

Figure 2.2 Sustainability as increasing per capita capital stock (four kinds of capital)

Source: Serageldin, I. (1996), *Sustainability and the Wealth of Nations, First Steps in an Ongoing Journey,* World Bank, Washington, DC, p. 4.

By addressing the application of this concept in terms of measurement and monetary valuation, the World Bank team faced formidable problems. It became very clear that our knowledge about the relationships between social structures, human activities and ecological processes is still fragmentary. While the identification of social capital turned out to be impossible for the time being, the World Bank actually calculated three kinds of capital, i.e., produced assets, natural capital and human resources. Without going into the

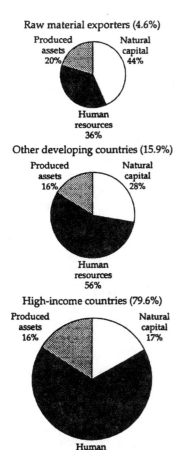

Raw material exporters (4.6%)

Produced assets 20%

Natural capital 44%

Human resources 36%

Other developing countries (15.9%)

Produced assets 16%

Natural capital 28%

Human resources 56%

High-income countries (79.6%)

Produced assets 16%

Natural capital 17%

Human resources 67%

Figure 2.3 Composition of World Wealth

Source: World Bank (1995), *Monitoring Environmental Progress*, World Bank, Washington, DC, p. 63.

methodological details of the – rather rough and arbitrary – calculation, the composition of 'world wealth' reveals that, with the exception of some raw material exporting countries, the value of human resources as defined by the World Bank equals or exceeds the aggregate value of both natural capital and produced assets.[1] Equally important is that produced assets represent only 16 to 20% of the wealth of most countries studied – as can be seen from Figure 2.3.

The policy implications of these findings are evident. They could be summarized as encouraging the growth of natural capital by reducing our level of current exploitation and, most importantly, increasing investment in human resources, particularly those of the poor, who are both the victims and the unwitting agents of environmental degradation in many of the earth's poorest countries.

Nepal: Unsustainable Development in one of the Earth's Poorest Countries

Nepal undoubtedly constitutes one of the most prominent cases of endangered ecosystems in the world. Bohle and Adhikari (1998) provide a large number of indicators that clearly point out the high and increasing degree of ecological, economic and social unsustainability prevalent throughout the country. It could be seen from data as given in Table 2.1 that Nepal in the South Asian context is by far the lowest ranking country in terms of the new World Bank index on the 'wealth of nations'. Though, with regard to natural capital, being only on the second last rank in South Asia (before Bangladesh), Nepal figures by far the last in terms of human capital. Even on a global scale, the country was ranked second last among 133 countries analysed (with only Ethiopia ranking worse).

Table 2.1   South Asia – World Bank index on the wealth of nations (capital in US$ per capita)

	Natural Capital	Produced Assets	Human Capital	Total
Sri Lanka	560	1,600	7,240	9,400
Pakistan	410	870	5,500	6,780
India	475	1,075	2,750	4,300
Bangla Desh	220	430	2,450	3,100
Nepal	270	430	900	1,600
Australia*	660,000	60,000	175,000	895,000
Ethiopia*	540	300	560	1,400

* First and last ranked countries worldwide.

Source: Fues, T. (1996), 'Humankapital und Naturvermögen: Der neue Weltbank-Index für Wohlstand und Nachhaltigkeit', *Entwicklung und Zusammenarbeit*, vol. 37, pp. 308-9.

Land productivity in Nepal stagnates inspite of increasing uses of fertilizer what could be interpreted as a clear indication of land degradation and excessive utilization of natural resources (Koirala, 1992). More than half of Nepal's population (60%) have to spend more than two thirds of their household budgets for food alone, while more than 80% of the food supply consists of cereals. The intensity of poverty is on a dramatic increase. While 40% of the population were below the poverty line in 1975/76, i.e., around 5.5 million people, the proportion increased to 43% in 1984/85 with a further rise to 49% in 1992/93, i.e., around 9 million people (Guru-Gharana, 1995 a, b; Government of Nepal, 1995).

During the last twenty years, Nepal has changed from a net exporter to a net importer of food (Cameron, 1995). The Food Balance Sheets for Nepal as provided by the Food and Agriculture Organization of the United Nations (FAO) reveal that since 1991/92 all years have been deficient – with the deficit increasing from year to year (Central Bureau of Statistics, 1995). In a scenario developed for Nepal by the International Development Research Centre's (IDRC) Cooperative Research Program, it is expected that 33 out of a total of 75 districts will be food deficient by the year 2000, while only 7 districts will achieve food surplus. This will be a dramatic change as compared to 1981 when only 8 out of 75 districts were found to be food deficient, while 40 districts had achieved food surplus (Schreier, Brown, Schmidt, Kennedy, Wymann, Shah, Shresta, Nakarmi, Dongol and Pathak, 1990).

At present, the proportion of food deficit populations is especially critical in the Middle Mountains of Nepal where in 1992 nearly half of the population (47%) was found to be undersupplied (Koirala, 1992). The respective figures for the lowlands along the Indian border (*Terai*) are 23% and for the High Mountains 31%. The still favourable situation in the *Terai* will probably turn to the worse. Therefore, one of the case studies presented here provides evidence from the Middle Hills and another from the *Terai*, both representing the present and future problem areas of Nepal.

On the background of ecological, economic and social unsustainability as outlined, questions as follows will further be explored in more detail. First, what are the most critical regions with particular emphasis upon food security? Secondly, who are the most vulnerable groups? Thirdly, what are the main risk factors that threaten the livelihoods of vulnerable groups? Fourthly, how do vulnerable groups in rural Nepal cope with endangered ecosystems and unsustainable development, and how do they try to adapt to changing internal and external impacts? Finally, and most important, how (un)successful are their coping and survival strategies?

Coping Strategies in Social Science Research

The social science research on coping strategies concentrates on the behaviour of people or social groups under conditions of life-threatening risk, existential uncertainty and cyclical or permanent crisis proneness. A central aspect of such research is the question to what extent vulnerable groups can adapt to life-threatening risks. Empirical work has demonstrated that coping strategies are generally not a mere passive adaptation but constitute rather active, flexible and innovative responses to cope with risk, stress and shocks (Spittler, 1989). In many cases, coping strategies evolve as sequential processes in the form of a concatenation of problems and problem-solving efforts (Mortimore, 1989; de Garine and Harrison, 1988). From the perspective of social science, a number of elements central for coping are discussed such as the role of indigenous institutions (Huss-Ashmore and Katz, 1989), social cohesion and institutions to solve crises (Rau, 1991), social networks (Bryceson, 1990), and gender relations (Vaughan, 1987).

In recent times, and particularly emerging from the Indian discussion, research on coping strategies has developed the concept of 'sustainable livelihood security'. In this context, Blaikie, Cannon, Davis and Wisner (1994, p. 9) define 'livelihood' as

> the command an individual, family or other social group has over an income and/or bundles of resources that can be used or exchanged to satisfy the needs. This may involve information, cultural knowledge, social networks, legal rights as well as tools, land, or other physical resources.

Relating to the elements of 'security' and 'sustainability', a definition is given by Dahl (1993, p. 21-2) in that

> security indicates protection, assurance or secure conditions. The livelihood is sustainable if it can bear the weight of present activities for a long period.

It is the operational model of 'sustainable livelihood security' that seeks to grasp the opportunities and constraints that regulate access to the central elements of livelihood. In this context, Blaikie, Cannon, Davis and Wisner (1994) have developed the so-called 'access model' where social networks, relations of power and dependence, resources and stocks, household budgets and qualifications are presented. Another concept which concentrates on the central elements of livelihood is that of 'assets', which has been widely employed in one of the latest Human Development Reports (United Nations Development Programme, 1997). This concept not only considers economic, but also social, political, ecological, infrastructural and personal assets.

In summary, coping strategies under conditions of risk and uncertainty imply the integration of individual and communal strategies which are employed with the objective to mobilize livelihood resources (Friedmann, 1996), assets (Swift, 1989; Chen,

1991) and social opportunities (Drèze and Sen, 1995) and to secure them in a sustainable way. The geographical perspective focuses on the specific ecological and social context in which livelihood security is negotiated and contested (Chen, 1991), on the social actors with their specific opportunities and constraints, and on the 'social interface' (Schlottmann, 1998) where success or failure of coping strategies are finally determined.

Case Study

In order to focus upon the ecologically and economically most critical regions of Nepal, one case study area was selected in Nawalparasi District which is situated in the transition zone between the *Terai* and the Siwalik Hills, and another in Kaski District located in the transitional zone between Pokhara Basin, Middle Mountains and High Himal. In both study areas, three to five villages were selected to be intensively surveyed. The villages chosen represent most remote villages with highly fragile and endangered ecological settings under the (relative) conditions of market unaccessibility – see Figure 2.4.

From the research tools applied in the four villages selected from the nothern case study area of Kaski District, Rapid Rural Appraisal (RRA) methods were used in Siklis, while Participatory Rural Appraisal (PRA) was carried out in Siding and Karuwa-Kapuche, with the village of Lachok having undergone a detailed survey of agrarian change (Adhikari, 1995). A similar research design was used for the southern case study area of Nawalparasi District, while data on 'wealth ranking' of more than 5,000 households in 1992 to 1996 were derived from unpublished materials provided by Action Aid Nepal.

Extent and Distribution of Food Deficit

In Table 2.2, the total of 356 households surveyed in the sample villages Siding and Karuwa-Kapuche of Kaski District are classified according to five groups of food self-sufficiency (FSS) using PRA-methods (Pyakuryal, 1995 a). In the case of Kaski, it has been found that 41% of the households were less than six months self-sufficient from own food production, while 51% were self-sufficient less than 12 months. Only 8% of the households were found to have food surpluses.

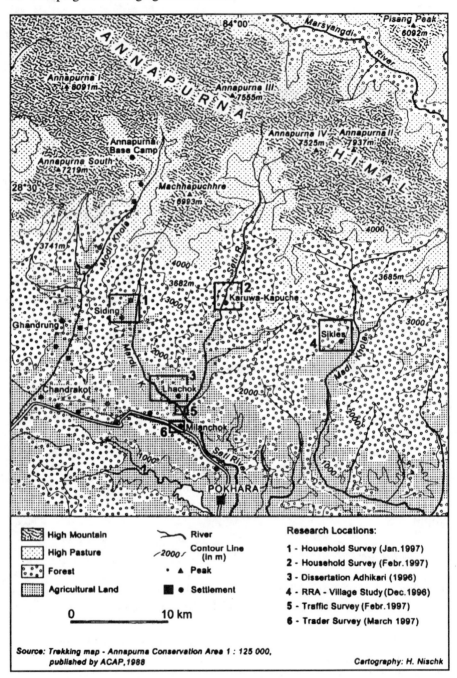

Figure 2.4   Location of Kaski case studies

Table 2.2   Classification and extent of food self-sufficiency (FSS) in two villages of Kaski District, 1996/97

FSS groups	Total number of households	... as % of all households	FSS per group
0-3 months	26	14	
4-6 months	52	27	41%
7-9 months	48	25	
10-12 months	49	26	59%
> 12 months	15	8	
Total	190	100	

Table 2.3 provides data from four sample villages in Nawalparasi District. It could be seen that the situation in food self-sufficiency was even worse when compared to the sample coming from Kaski District. In Nawalparasi, 68% of all households surveyed could not supply themselves for at least six months in a year, while 31% were food self-sufficient between 7 to 12 months with only 1% of the households having food surpluses.

Table 2.3   Classification and extent of food self-sufficiency (FSS) in four villages of Nawalparasi District, 1994/95

FSS groups	Total number of households	... as % of all households	FSS per group
0-3 months	37	22	
4-6 months	78	46	68%
7-9 months	39	24	
10-12 months	11	7	32%
> 12 months	1	1	
Total	166	100	

Source: Action Aid Nepal (1994/95), *Rural Livelihood Survey*, No. PA-2, Nawalparasi (unpublished data).

Food production in the villages surveyed came basically from three agricultural sub-systems, i.e., irrigated land (with 2 to 3 harvests per year, mainly paddy), rainfed land (with one harvest per year, mainly maize or millet), and slash and burn agriculture on the hill slopes (with one harvest every three years, mainly millet) (Pyakuryal, 1995 b). The higher the location of the villages, the more important are the latter two agricultural sub-systems for food production.

Table 2.4   Sources and consumption of food in the village of Siding, 1996/97

FSS	Required*	Produced	Purchased	Sold	Bartered	Consumed
0-3	100%	21%	50%			71%
4-6	100%	44%	31%		10%	85%
7-9	100%	51%	23%		2%	76%
10-12	100%	78%	19%	2%	2%	97%
>12	100%	127%	14%	40%		101%
Mean	100%	62%	25%	4%	3%	86%

* WHO standard of 180 kg of cereals per person and year.

Sources of Food and Consumption Levels

When calculating the level of consumption, the minimum food requirement has been assumed to amount to 180 kilograms of cereals per person and year, which is a standard value provided by the World Health Organization of the United Nations (WHO) – also being an absolute minimum amount as compared to other estimates. While the value as specified will correspond to about 1,650 calories per day and capita, the Nepalese government assumes that 2,250 calories could be held a suitable figure for defining the national poverty line (National Planning Commission, 1993).

In Table 2.4, it could be seen that in Siding village the lowest FSS-category produced only 21% of the minimum requirement (according to WHO), while another 50% were purchased. The actual consumption level was only 71% of the above defined minimum requirement. In the second FSS-category, the overall consumption level was 85%, in the third 76% and in the fourth 97% of the minimum requirement. On an average, only 86% of minimum food requirements in Siding were found to be fulfilled.

Data as given in Table 2.5 show that the situation in Karuwa-Kapuche village was even worse. In the lowest FFS-category, only 61% of the food requirement could be achieved, and the average for all households amounted only 82%.

Table 2.5   Sources and consumption of food in the village of Karuwa-Kapuche, 1996/97

FSS	Required*	Produced	Purchased	Sold	Bartered	Consumed
0-3	100%	16%	45%			61%
4-6	100%	39%	45%		7%	80%
7-9	100%	57%	31%		2%	75%
10-12	100%	88%	31%			118%
>12	100%					
Mean	100%	47%	41%		5%	82%

* WHO standard of 180 kg of cereals per person and year.

Summarizing the results, it could be stated that – in addition to own production – purchase and bartering of food are the main strategies to cope with food deficits, and that another coping mechanism being forced upon the people is to consume less than the minimum requirement needed for an active and healthy life. Thus, it is not surprising that apparently large parts of the mountain population are affected by stunted growth, although most of them are highly active and working hard.[2]

Sources of Income

What could be derived from the analysis of food sources and consumption levels is that most of the food deficit is covered by purchasing food in the market – see Tables 2.4 and 2.5. One source of income is wage employment in agriculture, in quarries and, most importantly, in portering services – see Figure 2.5. Alternatively, mountain produce is taken to the market such as timber, livestock, bamboo and also alcohol. These are all products for which remote mountain villages have comparative advantages. Extensive grazing grounds are available in high pastures or forest

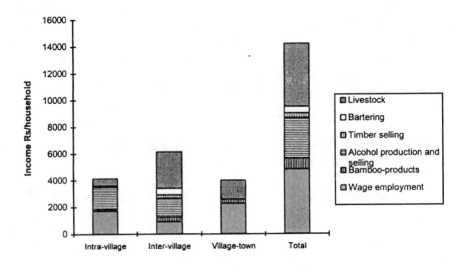

Figure 2.5   Income from various types of exchanges in Karuwa-Kapuche Village in 1996

mountains, so that elaborate transhumance systems have evolved. Timber, illegally cut in mountain forests, is sold in the market during night time in order to evade ranger controls (while any illegal access to the forests will be prosecuted). Being illegal, too, alcohol production is carried out requiring considerable amounts of firewood for the distilling process. Production and marketing is mainly undertaken by non-Hindu women who sell the alcohol in lower lying villages where they buy millet being an essential ingredient for the production of alcohol. Collected in the forests, bamboo is another income generating produce in the form of baskets and mats produced in the villages and marketed in Milanchok or Pokhara.

Spatial Structure of Market Linkages

In Figures 2.6 and 2.7, evidence is provided how the two remote villages surveyed are linked to the outside world by means of commodity flows. A difference is made between intra-village, rural-rural and rural-urban interactions – as could also be seen in Figures 2.5 and 2.8.

As a matter of fact, the high importance of inter-village linkages as compared to intra-village and rural-urban exchanges has to be noted first. In Karuwa-Kapuche, inter-village exchange accounts for 43% of the income sources of households surveyed, whereas intra-village exchange

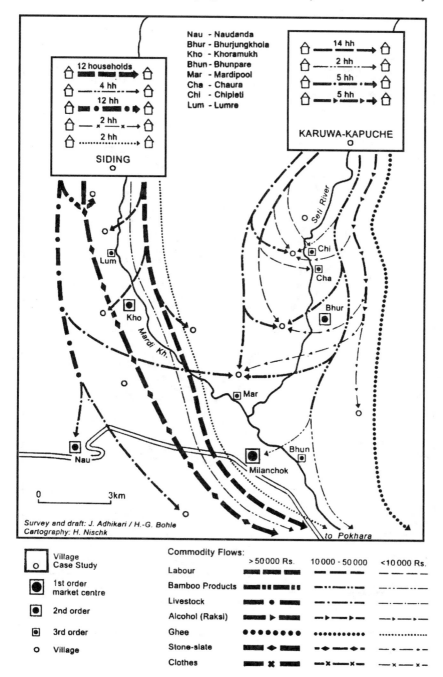

Figure 2.6   Outflows from Siding and Karuwa-Kapuche Villages (1996)

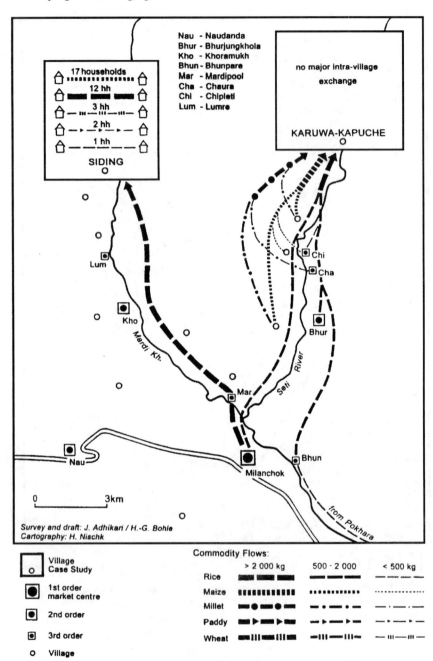

Figure 2.7   Inflows to Siding and Karuwa-Kapuche Villages (1996)

provides 29% and village-town linkages 28% of all the income. In the inter-village exchange network, livestock by far figures first, thus being a clear indicator of the comparative advantage of peripheral high mountain villages in terms of pastures and fodder. Alcohol production and selling, an activity which is highly dependent on high inputs of firewood, ranks second. In Siding, intra-village exchange dominates the income pattern (51%) but inter-village exchange contributes to 31%, with livestock being the most important activity. Income from village-town linkages, mainly derived from wage employment, was found to amount to only 18%. The findings emphasize the high relevance and complexity of inter-village exchange. They indicate that this sphere of exchange, as it was pointed out in Adhikari's (1995) survey of Lachok and Reban villages, is an important, but widely neglected and underresearched element of rural livelihoods in the villages of western Nepal.

Another point to be stressed is the high degree of spatial interaction, with labor and stone slates being the main export products from Siding, and alcohol, butterfat (*ghee*) and livestock being crucial for Karuwa-Kapuche. With regard to the pattern of inflows, rice is by far the most important commodity. This is also revealed by the trader survey where rice is ranked first among the commodities sold to the village people, followed by salt, sugar, vegetable oil, fresh vegetable, tea, meat, chillie and chowmein. Since chowmein is an instant soup which can be cooked within a minute being increasingly used by porters and laborers to have a quick meal *en route*, it gives a revealing example how new consumption patterns become to assist in saving firewood and time.

Figure 2.8   Income from various types of exchanges in Siding Village in 1996

Conclusions

As laid down by Dahl (1993), sustainable livelihood security has to be determined by standards of ecological security, social equity and economic efficiency. At present, it is attempted to develop a new 'Sustainable Livelihood Security Index' along these three determinants in the context of the Indian discussion (Swaminathan Research Foundation, 1992). The working hypothesis as it had been used here, i.e., that sustainable livelihood security is first of all determined by food self-sufficiency, could not be confirmed. The structure of vulnerability is found to be much more complex than being based on subsistence level alone. This becomes clear when data are disaggregated to the level of the individual household. In the case of Karuwa-Kapuche, a tendency is observed in that households with low food self-sufficiency are also highly deficit in total food consumption – what is graphically represented on the lower axis of Figure 2.9. However, in the upper categories of food self-sufficiency, deficits and surplus conditions occur without any significant correlation to the respective subsistence level. In these cases, the specific coping strategies of the individual household have to be examined. There are clear indications that success or failure of these coping strategies is determined by the size of the household following the rule 'the smaller, the more successful'.

Also, a clear tendency could be seen in that the resource basis of the individual household is determined by caste and ethnicity. Thus, in Figure 2.9 the point is made that it is not the sole exposure to risk (expressed in terms of subsistence levels), but that the main determinants of livelihood security and vulnerability could be identified as coping strategies being interwoven in a complex manner for each and every household.

By defining vulnerability, Chambers (1989) makes the difference of approaching or looking at it from two sides, i.e., an external side relating to exposure to stress, shocks and risks, and an internal side relating to the capacity of people of successfully coping with risks, stress and shocks. These two determinants of vulnerability clearly evolve from the study of livelihood security in rural Nepal as presented here. As a conclusion, it could be stated that access to life chances is decisively determined by the strategies which are adopted to cope with food deficits and uncertainty.

While the external side of vulnerability has mainly been examined from a macro-perspective of human ecology, entitlement theory and political economy (Watts and Bohle, 1993), the internal side of coping has to be examined from a micro-perspective. As one of the most striking research findings presented here, the point is made that unsustainable development   –   being   ecologically,   economically,   and   socially

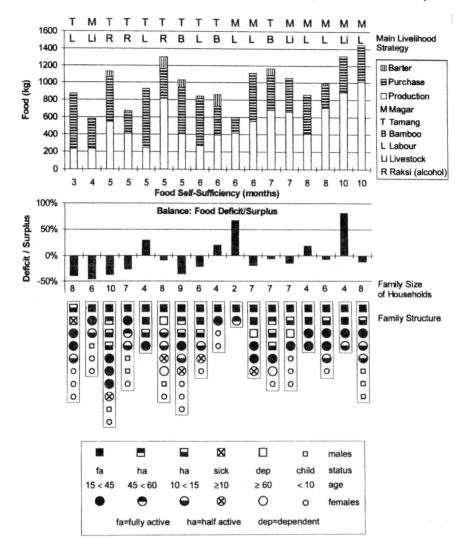

Figure 2.9 Livelihood profiles in Karuwa-Kapuche Village in 1996

determined – can only be made operational and visible on the local level, i.e., last but not least on the level of individual households. A micro-perspective of coping with vulnerability as adopted here will certainly constitute a domain of and challenge for human geography and anthropology.

Although one has to admit that for the analysis of the internal side of vulnerability a lack of theoretical approaches has to be noted, it has to be

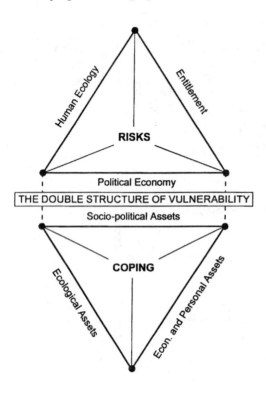

**Figure 2.10 The social structure of
vulnerability**

stressed that in the analysis of coping strategies used by vulnerable groups the internal side of vulnerability (i.e., coping with risk and factors determining success or failure) will be of pivotal importance. The focus to be selected here should be on the specific position of people or social groups within the total context of social relations as indicated in the concepts of 'access to resources' or 'assets'. Besides access to economic, social, political, personal and ecological assets, categories such as class, caste, ethnicity, gender and age will constitute basic variables of social vulnerability. A preliminary attempt to integrate such variables into an analytical framework of vulnerability is presented in Figure 2.10 in that a 'double structure of vulnerability' is identified that aims at integrating the external as well as internal dimensions while special emphasis is given to aspects of access and assets.

Notes

[1] For a world map of the 'wealth of nations', see Bohle, H.-G, and Graner, E. (1997), 'Arme Länder – Reiche Länder: Neue Untersuchungen über Nachhaltigkeit und den Reichtum der Nationen', *Geographische Rundschau*, vol. 49 (12), pp. 735-42.

[2] As a matter of fact, in one of the traffic surveys a man had been identified as having less than 1.5 metres in height (what is not uncommon in the area) and portering a bag of rice of 100 kilograms weight up the mountains – though he himself had a body weight of just 45 kilogram.

References

Adhikari, J. (1995), *The Beginnings of Agrarian Change, A Case Study in Central Nepal*, T.M. Publication, Kathmandu.

Blaikie, P., Cannon, T., Davis, J., Wisner, B. (1994), *At Risk: Natural Hazards, People's Vulnerability, and Disasters*, Routledge, London, New York.

Bohle, H.-G. and Adhikari, J. (1998), 'Rural Livelihoods at Risk: How Nepalese Farmers Cope with Food Insecurity', *Mountain Research and Development*, vol. 18 (4), pp. 321-32.

Bryceson, D. F. (1990), *Food Insecurity and the Social Division of Labour in Tanzania*, 1919-85, Macmillan, London.

Cameron, J. (1995), *Food Security: Background Technical Paper for the Agricultural Perspective Plan*, Government of Nepal, Kathmandu.

Central Bureau of Statistics (1995), *Statistical Year Book of Nepal 1995*, Government of Nepal, Kathmandu.

Chambers, R. (1989), 'Editorial Introduction: Vulnerability, Coping and Policy', *IDS Bulletin*, vol. 20 (2), pp. 1-7.

Chen, M. A. (1991), *Coping with Seasonality and Drought*, Sage, New Delhi, Newbury Park, London.

Dahl, S.-L. (1994), 'Sustainable Livelihood Security', *The Indian Geographical Journal*, vol. 68 (1), pp. 21-32.

Drèze, J. and Sen, A. K. (1995), *Economic Development and Social Opportunity*, Oxford University Press, New Delhi.

Friedmann, J. (1996), 'Rethinking Poverty: Empowerment and Citizen Rights', *International Social Science Journal*, vol. 148, pp. 161-72.

Garine, I. de and Harrison, G. A. (eds) (1988), *Coping with Uncertainty in Food Supply*, Oxford University Press, Oxford.

Government of Nepal (1995), *Nepal Agricultural Perspective Plan, Final Report*, Main Document, Government of Nepal, Kathmandu.

Guru-Gharana, K.K. (1995 a), *Human Development Strategy for Nepal: Perception from Below*, National Seminar on Development Strategy for Nepal: Perception from Below, National Planning Commission, Kathmandu.

Guru-Gharana, K.K. (1995 b), 'Trends and Issues in Poverty Alleviation in Nepal', *The Economic Journal of Nepal*, vol. 18 (1/69), pp. 1-16.

Huss-Ashmore, R. and Katz, S. H. (eds) (1989), *African Food Systems in Crisis, Part 1: Microperspectives: Food and Nutrition in History and Anthropology*, No. 7, Gordon & Breach, New York, London, Paris.

Koirala, G. (1992), *Proposed Approaches to Poverty Alleviation in Nepal*, National Seminar on Poverty Alleviation and Human Development, National Planning Commission, Kathmandu.

Mortimore, M. (1989), *Adapting to Drought: Farmers, Famines and Desertification in West Africa*, Cambridge University Press, Cambridge, New York, New Rochelle.

National Planning Commission (1993), *An Outline on Poverty Alleviation Policies and Programmes*, Government of Nepal, Kathmandu.

Pyakuryal, K.N. (1995 a), *Nepalese Farming Systems: A Socio-Cultural and Ethno-Historical Perspective, Background Technical Paper for the Agriculture Perspective Plan*, Agricultural Projects Services Centre, Kathmandu.

Pyakuryal, K.N. (1995 b), *Poverty in Nepal: Background Technical Paper for the Agricultural Perspective Plan*, Agricultural Projects Services Centre, Kathmandu.

Rau, B. (1991), *From Feast to Famine: Official Cures and Grassroots Remedies to Africa's Food Crisis*, Zed Books, London, New Jersey.

Schlottmann, A. (1998), *Entwicklungsprojekte als 'strategische Räume': Eine akteurs-orientierte Analyse von sozialen Schnittstellen am Beispiel eines ländlichen Entwicklungsprojektes in Tanzania*, Freiburg Studies in Development Geography No. 15, Verlag für Entwicklungspolitik, Saarbrücken.

Schreier, H., Brown, S., Schmidt, M, Kennedy, G., Wymann, S., Shah, P.B., Shresta, B., Nakarmi, G., Dongol, G. and Pathak, A. (1990), *Soils, Sediments, Erosion and Fertility in Nepal*, International Development Research Centre, Ottawa.

Serageldin, I. (1996), *Sustainability and the Wealth of Nations, First Steps in an Ongoing Journey*, World Bank, Washington, DC.

Serageldin, I. and Steer, A. (eds) (1994), *Making Development Sustainable: From Concepts to Action*, World Bank, Washington, DC.

Spittler, G. (1989), *Handeln in einer Hungerkrise: Tuaregnomaden und die große Dürre von 1984*, Westdeutscher Verlag, Opladen.

Swaminathan Research Foundation (ed) (1992), *Annual Report 1992*, M.S. Swaminathan Research Foundation, Madras.

Swift, J. (1989), 'Why are Rural People Vulnerable to Famine?', *IDS Bulletin*, vol. 20 (2), pp. 8-15.

United Nations Development Programme (1997), *Human Development Report*, Oxford University Press, New York.

Vaughan, M. (1987), *The Story of an African Famine: Gender and Famine in twentieth-century Malawi*, Cambridge University Press, Cambridge, New York, Port.

Watts, M. J. and Bohle, H.-G. (1993), 'The Space of Vulnerability: The Causal Structure of Hunger and Famine', *Progress in Human Geography*, vol. 17 (1), pp. 43-67.

3 Hydropower, Rice Farmers and the State

The case of deforestation in Laos

THOMAS KRINGS

Introduction

A typical process of deforestation could be identified if one takes the example of the Lao People's Democratic Republic (PDR) which together with Myanmar and Cambodia still has the largest forest cover of all the mainland countries in Southeast Asia (Talbott and Brown, 1998). According to official statistics, 47% of the country was covered by economically exploitable forests in 1995 as compared to about 70% in the 1960s. Out of political reasons, however, this figure seems to be 'improved'. According to Luther (1994), the actual share of forest cover in all land assumedly ranges between 30 and 40% in early 1990s. Particularly since the *Pathet Lao* troops took over government in 1975, a continuous decrease in the total area of forests has been observed. From 1975 to 1990, the annual rate of deforestation reached 1.4% as compared to 1.2% in 1990 to 1995 (Food and Agriculture Organization of the United Nations, 1997). It was noted by Hirsch (1995) that especially since 1985, Laos has experienced accelerated losses in forest cover.

Using the classification system of the United Nations, Laos belongs to the group of 'least developed countries' (LLDCs) by having a per capita income of US $380. About 85% of the 4.8 million inhabitants are engaged in the agricultural sector which still constitutes the backbone of the economy being widely marked by subsistence production. Moreover, 1.5 million people, with most of them living in the Northern provinces, still practise shifting cultivation, while about 60% of them are to be settled by the year 2000 according to government plans. In 1995, the agricultural sector contributed 56.3% to the national gross domestic product (GDP), while the industrial and service sector accounted for 18.8 and 24.9%. The agricultural sector faces several problems. Due to natural risks such as the flood hazards in 1995/96 and droughts in July 1998, for instance, severe problems of supply in rice (as staple food) have emerged in several regions of the country. The average yield of lowland rice, i.e., 3 tons per hectare, ranks among the lowest in South East Asia. Owing to an unsatisfactory infrastructure such as inadequate road facilities (about half of the population has no proper road access), the interregional exchange of rice proves to be difficult especially during the rainy season.

With an average population density of only 19.4 inhabitants/km², the country still is sparsely populated, but the present rate of population growth, i.e., 2.9% per year, turns out to be among the highest in all of Asia.

Since 1997, the rate of inflation ranks high as a result of the financial crisis having shaken the whole of the Asian region. Between June 1997 and June 1998, Lao people have experienced a loss of purchasing power capacity as high as 75% with the result of a turn to subsistence production – in rural as well as urban areas.

More than ever, Laos is a prime example of a 'donor-driven economy' (Evans, 1995; Pham, 1994) with external assistance amounting to US$ 60 per person and year. According to data of the International Monetary Fund (IMF), the foreign exchange reserves of the Central Bank of Laos declined drastically from US$ 143 million at the end of 1997 to US$ 122 million in autumn 1998. Taken together with the country's US$ 150 million trade deficit, this has adversely affected Lao Governments' potential for operating in a by now very tight fiscal environment.[1]

To conclude with an outline of Laos' developmental constraints, the country also suffers from being land-locked. Many of the nation's areas are physically isolated 'as a result of economic and social fragmentation, insufficient physical infrastructure, and high transportation costs, particularly for imports and exports' (Rigg, 1995).

Forest Resources

The forest cover of Laos falls into several types of natural forests. While monsoon or seasonal forests hold a significant extent of the total area under trees, at the end of the 20th century about 80% of the remaining natural forests assumedly will be degraded by logging or shifting cultivation. The most dominant type of forest is the 'mixed deciduous forest' which is found in hilly and upland areas between 150 and 800 metres above sea level receiving between 1,200 and 2,500 mm of annual precipitation. The characteristic species are *Tectona grandis*, *Pterocarpus* spp., *Terminalia* spp. and *Dalbergia* spp. Mixed deciduous forests comprise moist as well as dry types of 'mixed deciduous forest'. Due to a higher percentage of evergreen species and bamboo undergrowth, the moist deciduous forest more or less shows evergreen appearance. The dominant evergreen types of forest are the 'dry or semi-evergreen forests' (*Dipterocarpus alatus, Hopea ferrea, Anisoptera costata, Afzelia* spp.) and the 'hill evergreen forests' (*Cinnamomum* spp., *Fraxinus* spp., *Podocarpus* spp.) to be found in areas with annual rainfalls of more than 3,000 mm. In lowlands areas which receive less than 1,200 mm annual precipitation, 'dry Diptercarp forests' are widespread. They are characterized by open canopies and often by a low tree height (<8 m) with species such as *Diptercocarpus obtusifolius, D. intricatus, D. tuberculatus, Shorea obtusa.* Coniferous forests (*Pinus keysia, P. merkusii*) can be found on the plateaus above 800 m sea level in Northern and Central Laos (Hesmer, 1970; Foppes, 1995).

The tapping of natural resources, e.g., the exploitation of roundwood, cutbacks in mineral resources and the development of the hydropower sector, plays a vitally economic role in Laos' present state of development. Wood products account for 35 to 45% of the total foreign exchange earnings, while the share of forestry in the gross national product (GNP) is estimated to be 16% (Rigg and Jerndal, 1996). The point has to be made here that about 80% of domestic energy consumption remains wood-based. Forests also provide a host of non-timber forest products, including wild-life resources, mushrooms and medical plants, and a lot of high valued environmental functions which indirectly support agricultural and energy production (Khampheuane, 1998).

In the mixed deciduous forests, only a few economically precious woods such as ebony and rosewood (*Diospyrus* spp., *Dalbergia* spp.) are found. Wood of commercial value is commonly of only medium quality, mainly being vast amounts of *Dipterocarp* species (usage of which is also made in resin production). The most important woods for export being used in construction and for parquet flooring are *Pentacme, Shorea, Parashorea, Hopea* and *Afzelia*. The only natural teak forests left (*Tectona grandis*) are located near the border to Myanmar in the provinces of Oudomsay and Sayabouri. Species such as *Lagerstroemia spp.* are consumed for charcoal production especially by local populations. Due to an increase of secondary vegetation, Laos is covered by vast areas of bamboo forests which play an important role for local house construction and handicrafts. The commercial cultivation of rattan (*Calamus rotang L.*) is considered to create further economic potentials to be exploited in future (Hesmer, 1970; Foppes, 1995).

Political Ecology of Deforestation: An Actor-Oriented Approach

As in most tropical regions, the causes of national deforestation in Laos are very complex and have to be considered along the lines of regional differentiation. According to detailed reports as given by Brauns and Scholz (1997) on Southeast Asia and by Rigg and Jerndal (1996) in particular on Laos, different groups, institutions and users acting on different levels often follow quite distinctive (sometimes even contradicting) aims with regard to the usage of and impact upon forest resources. Particularly international and national economic interests prove to play a crucial role in the environmental degradation of natural forests.

According to the analytical framework of political ecology as outlined in Figure 3.1, the conflicting interests in forest resources are articulated by 'non-place-based-actors' (e.g., World Bank, Asian Development Bank,

Actors and their interests with regard to tropical forests from the viewpoint of Political ecology: *The example of Laos*			
	Actors	**Interests**	**Impacts on the tropical forest**
International level	Foreign Investors Mekong River Commision	Hydro-dam projects	Extensive clearing in future flooded areas of hydro-reservoirs
	Asian Development Bank	Road construction projects; hydroelectricity	Linear clearing and spontaneous settlement processes along new roads; extensive clearing of forests
	Foreign wood extraction Concessionaires	Timber production	Selective clearing of commercially valuable trees; decrease of valuable species
National level	Ministry of Economics Ministry of Forestry State Forestry Department Army	Till 1991: Exploitation of timber for foreign exchange earnings; establishment of commercial wood plantations Since 1991: development of hydro-power	Complete clearing of forests in future hydro-reservoirs
	Non-Governmental Organizations (NGO's)	Agro-forestry projects; Spreading of rural teak-plantations	Change of natural forests to commercially used forests; reforestation
Local level	Rural Population: shifting cultivators and wet-rice farmers	Securing of livelihoods and extension of market production sale of bamboo; hunting, use of non-timber forest products	Selective and extensive clearings due to shifting cultivation; decrease of species by charcoal production
			Th. Krings (1998)

Figure 3.1   Actors and interests with regard to tropical forests from the viewpoint of political ecology: The example of Laos

international hydro-power development companies) and by 'place-based-actors' such as governmental institutions, shifting cultivators, lowland rice farmers, hunters and gatherers. Within a multi-layered analysis, the strategies of different actors from the international, national and local level as well as the interests linking the actors from these levels will have to be included in any investigation aimed at the proper analysis of deforestation.

Thus, the dynamics of deforestation in Laos could be understood as a result of conflicting interests between groups who often act in the context of unequal power relations (Bryant and Bailey, 1997). Deforestation as in the special case of Laos emerges as the product both of modernisation (i.e., construction of large resource destroying infrastructural development projects such as hydropower dams) and increasing underdevelopment explained by indicators such as poverty level and high rates of population growth in combination with land shortages (Hirsch, 1995).

Actors on the Regional and Local Levels

According to the Government of Laos, shifting cultivation is the main driver of deforestation and forest transformation impacting adversely upon biodiversity and the environment. While an estimated 4.9 million hectares fell under the mode of shifting cultivation in 1989 (Chazee, 1994), about 150,000 hectares of forest fallow are cleared annually (Dufumier, 1996).

The regions with highest rates of deforestation are situated in northern highland areas such as in the provinces of Luang Namtha, Oudomsay, Sayabouri, Luang Prabang, Xieng Khouang and Houaphan. Considered to be mainly caused by extensive shifting cultivation as done by the Lao Sung (highland minority groups) and the Lao Theung (midland or upland Lao groups), the share of undisturbed forests there amounts to less than 20% (Rigg and Jerndal, 1996). Especially in the Province of Luang Prabang, large areas at an altitude of more than 800 m have already been completely cleared. As a result of land shortage, shifting cultivation becomes increasingly replaced by a system of permanent cropping on marginal lands. On slopes having a gradient of more than 40%, mountain dwellers such as the Hmong or Lao Theung cultivate upland rice and the undemanding *Andropogonoidae* grain-species called the 'tears of Job adlay' (*Coix lacryma-jobi L.*) and suitable to be grown on degraded soils. Soil erosion and land degradation, high variation of rice-yields and the lack of drinking water especially during the dry season are the main threats mountain dwellers are exposed to.

Completely cleared are the surroundings of the capital Vientiane (Vientiane Préfecture) and the areas along the Mekong river which traditionally are the most important rice producing areas in Laos.

The regions with a relatively high share of undisturbed forests are located in the Annamite Cordillera of Central Laos, in the provinces of Bolikhamsay and Khammouane. A mosaic of cultivated and forested land exists in the provinces of Savannakhet, Salavan, Sekong, Champasak and Attapu. As a general tendency, wooded areas being still intact could be found to increase from north to south – see Figure 3.2.

It has to be stressed that the on-going adherence to shifting cultivation and the lack of sufficient lowlands for wet-rice production reflect degrees of extreme poverty. Rural communities are exposed to risks such as climatic uncertainty, soil depletion and food insecurity. For that reason, intensive wildlife hunting and the illegal trade of protected animals and of wood products gain special significance in order to achieve livelihood security.

Although there is reason for being cautiously optimistic about Laos retaining representative samples of primary forests within its 17 National Biodiversity Conservation Areas (NBCAs), the prospects for sizeable populations of key wildlife and plant species are less certain. The risk that Laos will end up protecting 'empty and silent forests' (Khampeuane, 1998) is becoming more and more reality. The extinction of wildlife is going on in many areas because of the nationwide dissemination of hunting guns. More than two million hunting guns are estimated to remain in the hands of the Lao population. Most of the guns used could be found in the northern provinces where the proportion of poor rural populations is extremely high. There, nearly each household depends upon wildlife hunting for the adequate supply of family members with food.[2] Among the most endangered species are the giant *muntjak*, *saola* (a kind of forest antilope), mouse deers and various species of lizards and gibbons. Another source of income is cross-border trafficking of living animals including tigers, lizards, tortoises, Malaysian pangolins and many bird species some of which enter the medicinal production circuit of Chinese markets. Most intensive trafficking is reported from the Lao-Vietnamese border in Bolikhamsay Province.[3]

The State as a Key Actor in Forestry

In 1986, the Lao Government introduced 'New Economic Mechanisms' (NEM) as a policy initiative to promote market-oriented reforms (Saignasith, 1997). The opening up of the country to foreign investors led

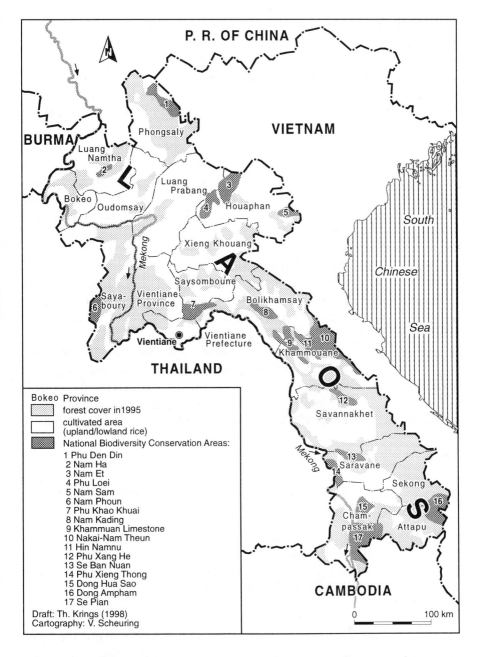

Figure 3.2   Forest cover and National Biodiversity Conservation Areas in Laos, 1995

to a rapid increase in logging and timber exports mainly to Thailand and Vietnam. In the late 1980s, 120 new joint ventures in the timber industry were signed between foreign companies and Lao partners. From 1986 to 1996, the volume of logged forests reached 400,000 to 500,000 cubic metres per year. Over 2.5 million hectares of forest were allocated to nine State Forest Enterprises with the aim of using wood resources as a basis for industrial development (Khampheuane, 1998; Rigg, 1995; Rigg, 1996).

Due to the increases in wood extraction, concerns about the future of the remaining forest resources in Laos were raised both in the country itself and in some Western donor countries. In 1991, the Government of Laos declared an official logging ban under the decree no. 67, thus consolidating central government control over the forestry sector and limiting the provinces' rights to issue logging concessions. Nevertheless, since 1991, the rate of logging and timber exports has continued to grow. The clearing of forests and felling of trees has been particularly permitted for hydropower sites to be built in future such as Nam Theun 2 in the province of Khammouane and Sekaman 1 reservoir in Attapu province. In these cases, long-term logging contracts had been given to companies from Thailand, South-Korea and Malaysia in 1992 (Daorung, 1997).

As one of the goals tried to be achieved by the Lao Government, higher increases in the value of the wood-processing sector are sought for since 1990. For that reason, the government signed up a joint-venture with a Malaysian enterprise in 1997 in order to establish a plywood and parquet producing factory at Ban Min which is about 100 km north of Vientiane.

Non-State Actors in Forest Destruction

About the impact of non-state actors in forest destruction it was noted by Bryant and Wilson (1998, p. 326) that

> international financial institutions derive power and influence as a result of their loan provision policies that may be linked directly or indirectly to the perpetuation of environmental management practices by other actors that may have critical implications for the destruction of human-environmental interaction generally.

In Laos, numerous infrastructural projects such as the improvement of roads and hydropower facilities are partly (or totally) financed by the Asian Development Bank (ADB). ADB's operations have expanded significantly with loan commitments averaging at approximately US$ 90 million per year with an additional US$ 5 million of technical assistance grants. Thus, ADB is one of the largest donors in the country.[4]

The negative impacts of road construction programmes are immediately visible in various parts of Laos. Between 1995 and 1998, many sections of national roads were sealed with the help of the People's Republic of China and Vietnam, e.g., the 500 km section of road No. 13 between Vientiane and Savannakhet. Even before the start of the improvement programme, the road section between the capital and Luang Prabang was tarred in 1994/95 with the northern part to Pak Mong soon to be finished. Since the most recent past (autumn 1998), the road from Pakse to Ban Hatsaykhoun near the Cambodian border is under construction at full stretch.

The environmental impacts of the often overhastily constructed roads are sometimes disastrous. This can be observed in Northern Laos between Vang Viang and Luang Prabang. On the extremely steep and totally denuded mountain slopes, cost-intensive side measures of road side development were neglected with the result that landslides are common blocking any transport during the rainy season. Since road construction rubble covers large areas along the road, valuable arable land gets lost for mountain-dwellers. From the very moment of road development, a rise in the number of settlements was the result. Many Hmong villagers who used to live on inaccessible mountain ridges started to settle down along the roads. One of the reasons behind was that vital water supply was guaranteed. At minor water-falls, drinking water is diverted by plastic tubes and road ditches serve as guiding rails for tubes taking water to the next road-side settlement.

In the context of road development, in the capital of Vientiane main roads are about to be enlarged and tarred since 1997. Simultaneously, many trees along the avenues planted during the French colonial times are felled, thus contributing to a detrimental change of micro-climate conditions such as loss of shadowed grounds and rise in surface temperature. The extension and modernisation of the traffic road system in the capital area is closely connected with the 'Visit Laos Year' in 1999 when up to 700,000 foreign tourists are expected to come to Laos.

Hydropower Project Developers

During the 1990s, plans to create a strong hydroelectric sector had been worked out. The extremely liberal foreign investment policy introduced in the course of economic reforms since 1989 has created a climate very much in favour of big foreign dam-building companies. By the year 2010, it is planned to have constructed more than one dozen hydropower projects by then.

According to Usher (1996) there is a 'race' for hydro, economic and political power in Laos involving country based actors as well as foreign investors mainly from Scandinavian countries. According to official estimates, Laos has an electricity generating potential that ranges between 18,000 and 24,000 mega watt if only the tributary rivers of the Mekong were taken (Krings, 1996; Kraas, 1997; Ryder, 1997). The national key actor clearly is the Government of Laos that looks upon hydro-power development as quick means of generating foreign exchange revenues. International players are foreign companies, the 'dam-builders', northern based construction firms and suppliers of dam-related equipment such as turbines, generators and transmission lines. Since 1993, Laos is the scene of intense competition among firms from Norway, Sweden, Australia and France. Strategic alliances are formed between the state and international players linking the interests of the state agents, foreign donors, consultants and dam-builders. Another actor has to recognized in the Mekong River Commission (MRC) which is a transnational river authority representing the governments of the four lower riparian states Thailand, Laos, Vietnam and Cambodia. MRC coordinates watershed management activities as well as dam-building programmes (Usher, 1996; Usher, 1998).

The two first hydro-power projects were built in the 1970s and 1980s at the Nam Ngum River (150 mega watt) and at the Xe Set (45 mega watt) in Southern Laos, mainly generating power for the Thai market. The third project, the Theun-Hinboun, is a trans-basin diversion project located at the border between the provinces of Bolikhamsay and Khammouane having an installed capacity of 210 mega watt and opened on April 4, 1998. The dam was completed with financial assistance from the government of Norway, the Asian Development Bank (ADB) and other private sources. At an agreed export selling price in 1994 of US$ 0.043 per kilowatt hour, all electricity generated is sold to Thailand. Concerns about severe ecological impacts as formulated by Usher (1996) materialized as early as during the first months following the start of operation. There is reliably reported evidence of various harmful and project-related effects such as declines in fish catches downstream of Nam-Kading-River, of some villages being impacted by the loss of riverside vegetable gardens and of increasing transport difficulties (Shoemaker, 1998).

The proposed Nam Theun 2 (NT 2) hydropower project located on the Nakai Plateau in the heart of Khammouane Province, as depicted in Figure 3.3, has also attracted substantial controversy due to concerns about high economic costs and significant environmental as well as social impacts.

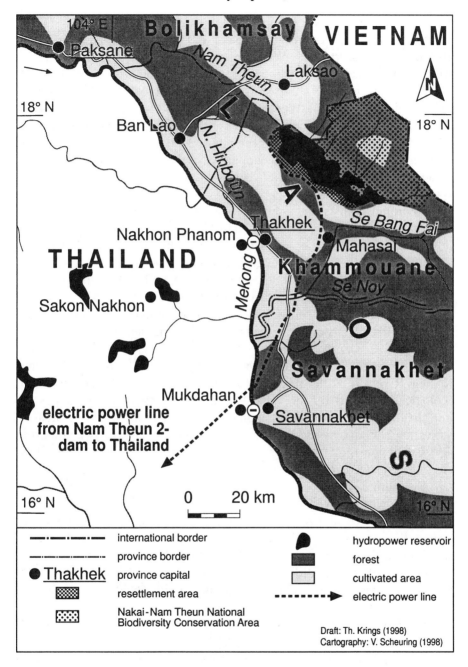

Figure 3.3   Location and spatial dimension of the planned Nam Theun 2 hydropower project in central Laos

Source: Bangkok Post, February 24, 1997.

When completed, the NT 2 reservoir will store approximately 3.18 million cubic meters of water covering an area of 450 square kilometers (i.e., about 40% of the Nakai Plateau). With a planned capacity of 681 mega watt, it will be one of the biggest hydropower projects in mainland Southeast Asia. The proponents of the NT 2 Electricity Consortium (NTEC) are Australia's Transfield-Company, the French energy-giant Electricité de France and a mixed group of Thai companies (Phatra Thanakhit, Jasmine and Italthai). Since 1995, the project has been subject to an extensive series of environmental, social and economic assessments by developers and independent experts working on behalf of the Lao Government and the World Bank.

Environmentalists and economists refer to the immediate detrimental effects on the downstream flow of Nam Theun River. Expecting a riparian flow of 2 m³/sec after the completion of the dam, this is considered to be too low to sustain fish or invertebrate species. The proposed dam with a height of 44 m and a crest length of 315 m will block upstream and downstream migration of the native fish population. The losses occurring in fisheries will mainly affect the livelihood systems of people down-stream. Further concerns are expressed in that the detrimental ecological effects of the recently opened Theun Hinboun project would add to the impacts of the NT 2 dam upon the total hydrological system of Theun-Kading River. If built, this would imply a number of economic and ecological problems for the local population and increased insecurity for rural livelihood systems by the inundation of scarce lowland rice areas in the valley bottoms, by the destruction of vegetable gardens (which during dry season constitute the basis of additional revenues), and by losses of virgin forests due to the construction of the reservoir (Wegner, 1997).

The future reservoir site as proposed covers about 400 square kilometres which have been covered with pines (*Pinus merkusii, Pinus keysia*) that were already cleared in 1995/96 in order to create a *fait accompli*. Forest losses will also occur due to the construction of transmission lines with strips of lands having a width of 150 meters where all trees have to be felled. Additionally, the resettlement of approximately 5,000 inhabitants living in the area of the proposed site could induce negative impacts upon forests outside the zone.

The Government of Laos, however, emphasizes that as an outcome of the project there will be assured funding for the management of the Nakai-Nam Theun National Biological Conservation Area (NBCA) for a period of 30 years. Nakai-Nam Theun NBCA forms the greater part of the project's catchment area and is unique in terms of biodiversity. Funding from electricity export revenues could preserve it from the stress induced by the construction of the reservoir and through the resettlement of about 5,000 people.

In anticipation of annual sales revenues of about US$ 250 million, Lao government puts strong emphasis on Nam Theun 2. NTEC officials are optimistic about delivering over 5,000 gigawatt hours of electricity per year to Thailand, and consultants estimated that the project would add an annual 3.2% to the gross national product of the country. Problems not yet solved, however, are the current pricing negotiations with the main buyer of electricity, the Electricity Generating Authority of Thailand (EGAT). With reference to the financial crisis, the Thailand Load Forecast Committee has sharply cut its power demand projection for the period 1997 to 2011. There remains the risk that Nam Theun 2 could be left without a buyer of electricity generated, or will be forced to sell electricity at a price too low to repay its huge construction costs of US$ 1,200 million.

Livelihood Systems in Sangthong District

Regional Setting

Taking the example of Sangthong District which is located about 60 km northwest of the capital Vientiane, livelihood systems will be examined against the background of increasing deforestation (Foppes, 1995; Thapa, 1998).[5]

Sangthong District is the most western district of Vientiane Municipality and is located in the northern vicinity of the Mekong River covering about 50,000 hectares of land. In geological terms, the Sayphou Phanang escarpment is part of the western fringe of the Vientiane Plain. The administrative boundary of the district follows the top of the escarpment, while in the west of it, the area is hilly and comprises moderately sloping hill ranges and small stream basins forming the watersheds of many tributaries of the Nam Thong and Nam Sang Rivers. To a considerable part, the hills are covered by secondary regrowth, bamboo forests (or shrub) and grasslands (Thapa, 1998).

According to author's own estimations, approximately 20,000 persons live there in 28 villages of the district. The average population density of 19 inhabitants/km² corresponds to the national average density. Most of the inhabitants are members of the ethnic group of Lao Loum. As a result of resettlements and in-migration, the ethnic structure has changed in some of the villages.

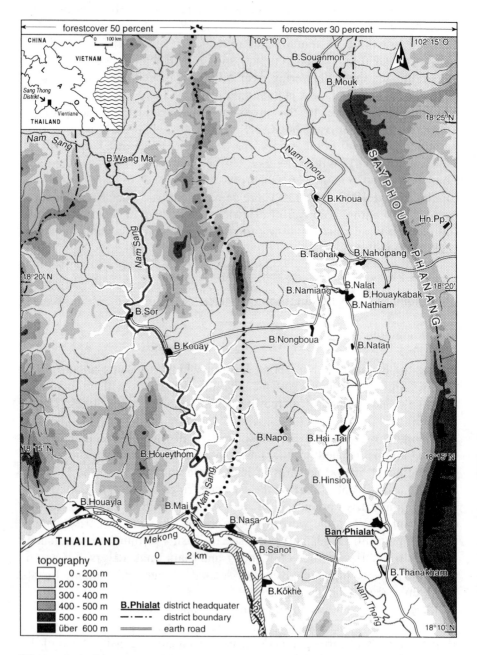

Figure 3.4   Location of Sangthong District (Vientiane Municipality)

As can be seen from Figure 3.4, the majority of the settlements are situated in the eastern part of Nam Thong River because of the availability of relatively flat land and a fair weather road linking the area with the district headquarter of Ban Phialat and the capital city of Vientiane. The population density of the western part of Sangthong is very thin with just a few villages located along the Nam Sang River.

Forest Transformation

To understand the process of deforestation, the district's environmental history will be touched briefly. In 1970, nearly 90% of the area was covered by virgin mixed *Dipterocarp* forests. Within a period of 25 years, nearly 60% of the forests had been converted into secondary forests or shrubland. The government, being the legal owner of the national forests, embarked on large scale commercial logging activities during the 1970s. In Sangthong, widespread logging started under a Japanese logging company (called Osaka) which harvested rosewood *(Pterocarpus macrocarpus)*, ebony *(Diospyrus rubra)* and *Afzelia* spp. for about three years. Osaka company was followed by the State logging company No. 9 operating in the area from 1981 to 1991. Within a period of ten years, two third of the natural forests disappeared (Thapa, 1998). Due to the construction of weather tracks to transport timber, many formerly inaccessible villages had been connected to the outer world. Following the proclamation of the logging ban in 1991, the activities of the State Company No. 9 were abandoned.

Besides commercial logging, other causes contribute to forest depletion. An important factor is the pressure on local villages to secure their livelihood. In some parts of the district, lowland rice production in seasonally flooded river plains is not sufficient to guarantee self-sufficiency in food so that additional land has to be cultivated for upland rice production under the system of shifting cultivation. As a further impact, in-migration into Sangthong District from other provinces of Laos and natural population growth during the past 20 years has induced and accelerated losses in forest cover (Thapa, 1998).

Present Systems of Land Use

The local livelihood system is based upon the cultivation of lowland rice (*naa*) as well as upland rice (*hai*). The oldest paddy fields to be found in the district are 200 years old. The opening of forests by logging companies in the 1980s led to a rapid increase of paddy fields. The construction of paddy fields is hard work which can take several years. To establish paddy fields, the villagers survey small streams and look for areas which are flat enough and will dispose of sufficient water nearby the settlements. In the first year, villagers cut most of the trees, burn the area and plant upland rice. During the weeding process, they continue to remove wood. In a next step, they have to dig paddy bunds for the retention of rain and run-off water. In areas with a slight gradient, future rice fields are divided into smaller units which get bunds along contour lines. The levelling of each field is done by repeated ploughing making use of buffaloes. It can take several years before all stumps are removed and the plot is really flat. The whole process is very labour intensive and may take five to ten years.

For socio-economic reasons, the cultivation of lowland rice has many advantages over upland rice production. Not only the labour requirements for weeding are much lower, but yields are higher and more stable. In the village of Ban Kouay, the average yield was 2,500 kg per hectare in lowland rice and 1,940 kg/ha in upland rice. Moreover, the risks of soil erosion are reduced in the case of lowland rice cultivation. Another advantage of lowland rice cultivation is the long-term conservation of soil fertility. In Sangthong District, buffaloes and cattle graze on paddy grounds after harvest at the same time manuring the paddy fields. Lowland rice can only be cultivated in the long run if forest cover on watersheds will be conserved. Paddy farmers therefore show a great interest in watershed protection in order to secure the delivery of water to their paddy fields.

Generally speaking, an increase of lowland rice areas, as it has been the case in the flat eastern parts of the district along the Nam Thong river, reduces the dependence on shifting cultivation. In Ban Taohai, where the farmers have a lot of paddy fields, upland rice cultivation has almost disappeared. If shifting cultivation is no longer practised, the fallow land may be reverted to secondary forest (Foppes, 1995).

In contrast to the eastern areas of the district, in the western part of Nam Sang River, shifting cultivation is still predominant in the village areas of Ban Kouay, Ban Sor and Ban Wang Ma since they lack land suitable enough to be grown with lowland rice. Shifting cultivation and the existence of fallow land is a prerequisite for the survival of wildlife, too. Fallow areas also provide significant amounts of firewood and animal fodder that can be produced there.

Livelihood Systems Compared

To understand the micro-regional differences of livelihood systems in the district, a total of 15 households in two villages of the eastern and western

Rice production, household consumption , surplus and deficits in 1994
(sample:15 households)

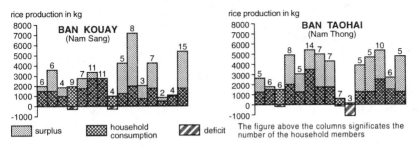

surplus   household consumption   deficit   The figure above the columns significates the number of the household members

Use of bamboo canes in 1994
(sample:15 households)

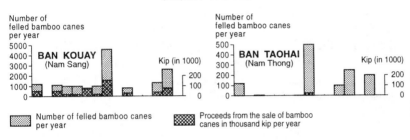

Number of felled bamboo canes per year   Proceeds from the sale of bamboo canes in thousand kip per year

Frequency of consumption of mouse deers (Tragulus javanicus)
und wild boars (Sus scrofa) in 1994
(sample:15 households)

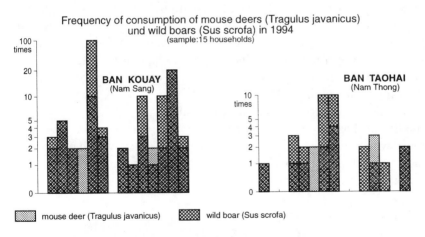

mouse deer (Tragulus javanicus)   wild boar (Sus scrofa)

Figure 3.5   Comparison of livelihood systems of Ban Kouay (Nam Sang) and Ban Taohai (Nam Thong) villages of Sang-thong District

Source: Foppes, J. (1995), *Farming and Forest Use Systems: Survey and Recommendations for Community Forestry in the Sang Thong Training and Model Forest*, Lao-German Forestry Team, GTZ Mission Report, Vientiane.

part were surveyed and compared, i.e., in terms of rice production, collection of bamboo canes and consumption of wild animals (*Tragulus javanicus, Sus scrofa*). As can be derived from Figure 3.5, the differences in livelihood strategies between Ban Kouay (village in the west of Nam Sang) and Ban Taohai (village in the east of Nam Thong) mainly reflect income possibilities in two ecologically distinct areas in 1994.

With regard to household consumption, a remarkable difference exists between the villages in terms of surplus and deficit production of rice. Quantities harvested and surplus achieved are by far larger in Ban Taohai where all families have access to paddy land as compared to Ban Kouay where only a few families possess paddy fields. In Ban Taohai, 60% of the families were able to sell rice, while in Ban Kouay the respective figure is only 22%. Rice production in Ban Kouay is based on shifting cultivation. Households, however, which suffered production deficits in 1994 could be found in the two of the villages. The lack of paddy fields in Ban Kouay and land shortage in Ban Taohai are identified to be the main reasons behind food insecurity. In Ban Kouay, 33% of the families have to buy rice, while the respective figure in Ban Taohai is only 20% (Foppes, 1995).

The use of bamboo canes for domestic purposes as well as for sale is much more prevalent among households in Ban Kouay area where still large tracts of secondary forest exist. In nutritional matters, the frequency and consumption of wild animals turns out play a major role among Ban Kouay people as compared to Ban Taohai. The remaining secondary forests in the western part of the district are the habitat of wild boars and mouse deers that are hunted not only for subsistence but also for the neighbouring Thai market. Other important sources of cash income constitute the sale of buffaloes, cattle and charcoal.

Finally, it has to be emphasized that the existence of degraded secondary woodlands is even the prerequisite for wildlife which constitutes a basic element for livelihood security in remote areas where rice production often turns out to be insufficient.

Towards Village Forest Management

In order to protect the district's secondary forests, it is suggested to put into practice a decentralised forest development and management strategy. This will only be possible through the active participation of the villagers. Since 1997, the Government of Laos has been developing a national village forest strategy the major objectives of which are to develop and promote village forestry as a means of sustainably managing Lao's forests (Khampheuane, 1998). According to Thapa (1998), however, the adaption of a participatory integrated forest management strategy in Laos will face several obstacles to be overcome. First, the government has not yet made efforts to institutionalise a participatory forest management system since the state is the exclusive and legal owner of natural resources. Any reform of property rights is suggested to aim at the legalisation of private ownership of agricultural and forested land and should be implemented in combination with a land register survey of all the villages' lands. The catastral survey should facilitate information on the land use, land managers, location and size of private land. Secondly, participative village forest programmes should take into consideration the practices of local resource management. In some villages, for example, so-called 'protected' or 'sacred' forests exist such as in the case of Ban Namiang where shifting cultivation is only practised scarcely. In general, village-based and self-reliant forest protection are seen to highlight the future potentials for forest management. Particularly paddy owners can be identified to have a strong interest in watershed protection in oder to secure the delivery of water to their lowland rice fields. Official forestry policies should therefore reach out to encourage those interests and to involve farmers in land use planning and watershed management (Foppes, 1995; Thapa, 1998; Krings, 1998).

Conclusions

Forest cover in Laos proves to be the object of many (and often contradicting) interests articulated by various actors. The most powerful actors with regard to forest use and forest transformation are surely to be found on the national level. Up to 1991, commercial logging activities which were favoured by the national government have contributed to substantial losses in forest cover inducing an increase of secondary forests. At present, and due to economic globalisation, international companies such as large energy plant suppliers are about to invest in the sector of hydropower in compliance with national development planning with negative impacts to be observed particularly upon the last forested areas in northern and central Laos.

If there would be a real official intention to protect remaining forests in future, then the key to a solution of environmental problems in terms of sustainability will be found at the local level where place-based groups of land users try to secure their livelihoods within and from the forests.

To achieve the goal of sustainable forest conservation and management, the concept of Village Forestry should be developed in the near future. According to the Department of Forestry, it is envisaged to attain the objective within a period of 15 years. In a first phase of implementation (1997-2002), main activities are planned as follows. First, strong policy and legal measures will be introduced such as Village Forestry Regulations to be drafted and issued. Second, national coordination of Village Forestry projects will be strengthened. Third, various models will be worked out in order to be applicable to different conditions with regard to forest (and other biological resources), cultural context and physical setting. And fourth, training capacities in Village Forestry will be strengthened (Khampheuane, 1998).

Notes

[1] Reported information as drawn from *Vientiane Times* (Laos), 1998; September 22-24, p. 1, September 25-28, p. 10, October 20-22, p. 11, October 30-November 1, p. 1, 20.

[2] See note 1.

[3] See note 1.

[4] See note 1.

[5] The data presented, if not specified otherwise, are based upon own field surveys done in Sangthong District of Laos in 1995.

References

Brauns, T. and Scholz, U. (1997), 'Shifting Cultivation: Krebsschaden aller Tropenländer?', *Geographische Rundschau*, vol. 49 (1), pp. 4-10.

Bryant, R. L. and Bailey S. (1997), *Third World Political Ecology*, Routledge, London.

Bryant, R. L. and Wilson, G. A. (1998), 'Rethinking Environmental Management', *Progress in Human Geography,* vol. 22 (3), pp. 321-43.

Chazee, L. (1994), 'Shifting Cultivation Practices in Laos: Present Systems and their Future', in D.V. Gansberghe and R. Pals (eds), *Shifting Cultivation Systems and Rural Development in Lao PDR, Report of the Nabong Technical Meeting*, United Nations Development Programme, Ministry of Agriculture and Forestry, Vientiane, pp.66-97.

Daorung, P. (1997), 'Community Forests in Lao PDR: The New Era of Participation', *Watershed*, vol. 3 (1), pp. 38-44.

Dufumier, M. (1996), 'Minorités Éthniques et Agriculture d'Abattis-Brûlis au Laos', *Cahiers des Sciences Humaines*, vol. 32 (1), pp. 195-208.

Evans, G. (1995), *Lao Peasants under Socialism and Post-Socialism*, Silkworm Books, Chiang Mai.

Food and Agriculture Organization of the United Nations (1997), *State of the World's Forests 1997,* FAO, Rome.

Foppes, J. (1995), *Farming and Forest Use Systems: Survey and Recommendations for Community Forestry in the Sang Thong Training and Model Forest*, Lao-German Forestry Team, GTZ Mission Report, Vientiane.

Hesmer, H. (1970), *Der kombinierte land- und forstwirtschaftliche Anbau: Tropisches und Subtropisches Asien*, Klett, Stuttgart.

Hirsch, P. (1995), 'Deforestation and Development in Comparative Perspective: Thailand, Laos, Vietnam', in O. Sandbukt (ed), *Management of Tropical Forests towards an Integrated Perspective*, Centre for Development, University of Oslo, pp. 37-50.

Khampheuane, K. (1998), 'Summary of the Country Outlook: Lao PDR', Asia-Pacific Forestry Sector Outlook Study, Working Paper No. APFSOS/WP/38, FAO, Rome.

Kraas, F. (1997), 'Instrumentalisierung des Mekong: Wasserkraft und fremdbestimmter Wirtschaftsaufschwung in Laos', in T. Hoffmann (ed), *Wasser in Asien, Elementare Konflikte*, Secolo, Osnabrück, pp. 364-68.

Krings, T. (1996), 'Politische Ökologie der Tropenwaldzerstörung in Laos', *Petermanns Geographische Mitteilungen*, vol. 140 (3), pp. 161-75.

Krings, T. (1998), 'Zerstörung der Tropenwälder: Ein globales Problem dargestellt am Beispiel von Laos', *Geographische Rundschau*, vol. 50 (5), pp. 291-98.

Luther, H.U. (1994), 'Laos' , in D. Nohlen and F. Nuscheler (eds), *Handbuch der Dritten Welt, Volume 7: Südasien und Südostasien,* Dietz, Bonn, pp. 436-56.

Pham, C. do (ed) (1994), *Economic Development in Lao PDR: Horizon 2000*, Vientiane.

Rigg, J. (1995), 'Managing Dependency in Reforming Economy: The Lao PDR', *Contemporary Southeast Asia*, vol. 17 (2), pp. 147-71.

Rigg, J. and Jerndal, R. (1996), 'Plenty in the Context of Scarcity: Forest Management in Laos', in M.J.G. Parnwell and R.L. Bryant (eds), *Environmental Change in South-East Asia*, Routledge, London, pp. 145-62.

Rigg, J. (1996), 'Uneven Development and (Re)Engagement of Laos', in C. Dixon, G. Drakakis and D. Smith (eds), *Uneven Development in South East Asia*, Ashgate, Aldershot, pp. 148-65.

Ryder, G. (1997), 'Stauen des Mekong: Regionale Energiepolitik', in T. Hoffmann (ed), *Wasser in Asien, Elementare Konflikte*, Secolo, Osnabrück, pp. 356-63.

Saignasith, C. (1997), 'Lao-style New Economic Mechanism', in M. Than and J.L.H. Tan (eds), *Lao's Dilemmas and Options*, Institute of Southeast Asian Studies, Singapore, pp. 23-47.

Shoemaker, B. (1998), *A Field Report on the Socio-Economic and Environmental Effects of the Nam Theun-Hinboun Hydropower Project in Laos*, International Rivers Network Report, Berkeley.

Talbott, K. and Brown, M. (1998), *Forest Plunder in Southeast Asia: An Environmental Security Nexus in Burma and Cambodia*, Environmental Change and Security Project Report No. 4, Woodrow Wilson Center, Washington.

Thapa, G.B. (1998), 'Issues in the Conservation and Management of Forests in Laos: The Case of Sang Thong District', *Singapore Journal of Tropical Geography*, vol. 19 (1), pp. 71-91.

Usher, A.D. (1996), 'The Race for Power in Laos: The Nordic Connections', in M.J.G. Parnwell and R.L. Bryant (eds), *Environmental Change in South-East Asia*, Routledge, London, pp. 145-62.

Usher, A.D. (1998), 'Dam as Aid: The Case of the Nordic Bilaterals', *Watershed*, vol. 3 (3), pp. 21-31.

Wegner, D.L. (1997), *Review Comments on 'Nam Theun 2 Hydroelectric Project Environmental Assessment and Management Plan (EAMP)'*, International Rivers Network Report, Berkeley.

4 Debating Vulnerability, Environment and Housing

The case of rural-urban migrants
in Cape Town, South Africa

BEATE LOHNERT

Introduction

Only as late as in the early 1990s, the role of cities in the developing world became part of the research agenda on global environmental change. The strategy paper on urban environmental management in developing countries, jointly published by the United Nations Development Programme (UNDP), the World Bank and the United Nations Centre for Human Settlements (UNCHS, HABITAT) in 1993 is the first document of international organisations explicitly stressing the important role of Third World cities within environmental change processes (UNDP, World Bank and UNCHS, 1993).

While the problematic increase of Third World urban populations has been broadly acknowledged, the working mechanisms, however, of rural-urban migratory system and their diverse environmental effects are hardly understood or documented. The lack of information about the problems and potentials of newly created marginalized livelihood systems in urban environments not seldom lead to policy failure. For applied research, this means bridging the divide of a sectorally and spatially independent analysis of the rural and the urban (Lohnert, 1998; Takoli, 1998).

Taking Cape Town as an example, this contribution focuses on rural-urban migrants and the problems they face in the context of the housing crisis in South Africa. Since the 1980s, South African cities are facing an immense in-migration of poor rural populations from the former *Homelands* which have been economically and socially neglected during *Apartheid* times. At the outskirts of the great cities informal squatter settlements are mushrooming to an extent and with rapidity that leaves very limited scope for urban planning concepts directed towards sustainable management.

In this context, very little is known about the motivation of migrants leaving their rural areas as well as the factors influencing the creation of new urban livelihood systems and its impact on urban ecology.

Rural-Urban Migration as Coping Strategy?

The theoretical framework of this contribution is based on the concept of social vulnerability. Watts and Bohle (1993, p. 118) define the three dimensions of vulnerability as

- the risk of exposure to crises, stress and shocks,
- the risk of inadequate capacities to cope with stress, crises and shocks, and
- the risk of severe consequences of, and the attendant risks of slow or limited recovery (resilience) from crises, risk and shocks.

A key issue that derives from the concept of vulnerability is the notion of coping strategies. The more diversified the portfolio of strategies is a person or group can dispose of to cope with vulnerability, the better are the chances to keep up the *status quo* or even enhance the respective living conditions. However, what is good for the individual at one point in time can have long term negative effects on the society as a whole. This could imply a drop back to the actors, hence, creating new dispositions for vulnerability. Thus, and particularly in the context of global environmental change, the issue has to be considered what and how are the long-term impacts of individual and group actor strategies.

In that sense, migration to urban centres constitutes a strategy to cope with environmental and socio-economic degradation in rural areas for the individual and his or her dependants. It can also reduce the pressure on ecosystems in areas where population size, resources and environment are no longer in a state of equilibrium. On the other hand, and as a general matter of fact, mainly the young men and women leave their rural homes, thus extracting the most productive people from marginalised rural areas and leaving the old and very young behind. As with processes of brain-drain, these people are no longer available for the development of their areas of origin.

With regard to urban areas, in-migration is a multi-facetted process. For the urban economy – be it formal or informal – migrants from rural areas form an important reservoir of cheap labour. At the same time most Third World cities are not able to provide adequate housing for the migrants.

Consequently, while looking for a better life, rural-urban migrants very often have no other choice than settling informally on urban ground that is not suitable for human settlements. For them, putting up a shack seems to be the only logical strategy to cope with the housing crisis. Lacking basic infrastructure such as water-supply, drainage systems and organised refuse disposal etc., informal settlements lead to a degradation and disappearance of urban open space and contribute to urban – and on an aggregated level, to global – environmental problems. In this context, migration as a coping strategy does not necessarily mean a betterment but may often result in long term deterioration of living conditions for the individual as well as for society as a whole. In this respect rural-urban migrants are responsible for and at the same time victims of processes of environmental change.

Housing as a Key Environmental Problem in Third World Cities

Environmental problems in cities occur in different contexts and have effects on different scales: from the smallest entity of the individual and household level up to municipality, city-region and a global level. This also applies to the housing problem in informal settlements.

The term squatter or informal settlement is used here to describe a form of housing that is created without formal authorisation and mostly initiated by the dwellers themselves. Most often these settlements are set up on free urban peripheral land, with or without legal title. The structure is dominated by single – very often extended – family housing that sometimes incorporates tenants and is characterised by a high percentage of owner-occupiers. Lacking or poor infrastructure and urban services are a result of the illegal status of these settlements. Housing quality is low and of makeshift nature, reflecting the low income levels of the population. As a category of low-income housing, the term squatter settlement covers a whole range of different physical qualities, locations and degrees of recognition by authorities (Mathéy, 1992; Obudho and Mhlanga, 1988; van Westen, 1995). In the following, some of the major causes and interlinkages of poor housing conditions and environmental problems are illustrated.

On the individual and household level, poor quality housing (such as makeshift houses, lacking the basic environmental infrastructure) impose a great health burden on the population residing there. Water-borne, airborne, and food borne diseases, as well as those which are passed on through insect or animal vectors constitute environmental factors as they are easily transmitted within a down-graded urban ecology and could – to a large extent – be reduced by a sustainable environmental city management. The economic consequences in terms of work days lost and costs of treatment due to the exposure to health risks reinforces vulnerability and poverty. In the absence of 'clean' energy provisions, open fires for cooking and heating lead to indoor air pollution and very often to extended fires, not seldom affecting total neighbourhoods. The inadequate protection by shelters of rain, wind and extreme temperatures put an additional strain on the health of the inhabitants.

On the neighbourhood level, high population density which is a common feature of informal settlements has reinforcing effects in terms of increasing the risk and rapidity of disease transmission and the spread of fires. Inadequate or non-existent sanitation forces inhabitants to defecate in the open, while insufficient drainage induces a spread of faecal contamination. In the absence of regular garbage collection, the burning of solid waste causes extensive air pollution. Additionally, most informal settlements are built on ground that is not suitable for human settlements, thus, enhancing their vulnerability to physical hazards like landslides and flooding. Very often, informal settlements are built on open ground along large traffic lines, resulting in an increased exposure of the inhabitants to traffic induced air pollution, accidents, and noise. Another problem, being widely observed in social milieus where people are exposed to a whole range of environmental stress, is the increase and intensification of social violence.

On the city and city-region level, surface, ground water and soil pollution through insufficient water and waste management are as prevalent as air pollution through increased traffic and industrial activities. Another problem faced by an unregulated increase of settlements is the disappearance of open ground in inner cities as well as at the periphery by the sealing of soil. Moreover, a loss of city-near agricultural land leads to enhanced transport needs.

In summary, the totality of problems mentioned here will finally aggregate and impact on processes of global environmental change in terms of exploiting non-renewable resources like fossil fuels and resulting in losses of biodiversity. They contribute to an overuse of finite renewable resources, e.g., freshwater, soil and flora, thus, adding to globally cumulative changes and they lead to an overuse of non-renewable sinks by the emission of persistent chemicals, greenhouse gases and stratospheric ozone depleting chemicals, hence, adding to globally systemic changes (Meyer and Turner, 1995).

Environmental changes on a global scale could only be understood if the social and economic driving forces on a local and individual level were made explicit. Thus, by shifting the perceptions, strategies and needs of rural-urban migrants in the centre of the analyses, two informal settlements in Cape

Town are presented here as examples. The following aspects which influence urban environmental change processes will be discussed:

- the importance of relations between the rural and the urban;
- the functioning of social networks and its impact on the urban environment; and
- the actual perception of problems experienced by the inhabitants of the informal settlements.

Cape Town as a Destination of Rural-Urban Migrants

Housing shortage constitutes a heritage of *Apartheid's* disparate development policy and is one of the main challenges the Government of the 'New South Africa' is facing. Segregation policy has created an urban spatial pattern – the so-called *Apartheid City* – that is unique in the world. Its socially painful aftermath will occupy South African urban planners and policy makers for decades. The fact that about 19% (approximately 8 million people) of the country's population live in informal or squatter settlements illustrates the scale of the problem.

Before *Apartheid Legislation*, especially before the *Group Areas Act*[1] regulating racially discriminating land rights and implemented in the beginning of the 1950s, Cape Town was said to be the least segregated city in South Africa. By 1985, however, Cape Town had become the most segregated major city in the country (Lemon, 1991) with the majority of the inhabitants classified as *Non-Whites* or *Africans* being evicted from the city far away to the outskirts of the town. The socio-economic pattern of Cape Town today still reflects this history with more than 50% of the township households living below the official household subsistence level (Metropolitan Spatial Development Framework, 1995).

In the process of restructuring, several severe constraints and problems will have to be tackled as outlined in short:

- high levels of poverty and unemployment, especially among those parts of the population that had to suffer under *Apartheid*-laws, with the overall rate of unemployment being as high as 36% according to different sources (MSDF, 1995);
- lack of basic infrastructure services (such as water, electricity and sanitation), inadequate transport system and unequal access to social infrastructure such as health care and education facilities;

- lack of adequate and affordable housing, particularly close to job opportunities: only 31% of the Cape Metropolitan Region's population is adequately housed, with an estimated 35% of the whole population living in informal squatter-settlements and 34% in so-called site and service schemes (MSDF, 1995);
- rapid and sustained natural population growth: according to conservative estimates, population growth averages 2.3% per year (Association for the Promotion of the Western Cape's Economic Growth, 1996).
- Rapid migration-induced population growth: it is estimated that 7,000-10,000 new informal settlers arrive per month;[2] from now about 3.2 million inhabitants, Cape Town is expected to reach the 5 million bench mark already in 2005 (MSDF, 1995).

Since the abolition of the *Influx Control Legislation* in 1986, Cape Town, as many other cities in South Africa, experienced a dramatic increase in the number of migrants. These originate generally from rural areas, especially from the areas of the former *Homelands* Transkei and Ciskei. This rural-urban migration resulted in a dramatic expansion of squatters and backyard shacks in the *African* townships, thus, increasing population density there to an extreme. At the same time, new-coming migrants and dwellers from the townships were increasingly creating new informal settlements on urban free land.

The unique flora of the Cape Peninsular which is classified in the highest biodiversity zone 10 (Barthlott, 1998) is severely endangered by an unsustainable sprawl of formal and informal housing. Especially the spreading of low-cost housing areas in the Cape Flats led to a decimation of the original flora by 94% (Boucher, 1983). Here, the loss of ground-water resources is particularly evident. The Cape Flats aquifer once had been one of the most important water resources in the area. Since 1975, the area has been sealed with low cost housing for persons formerly classified as *Non-White* and *African*.

The extensive urban sprawl of the Cape Metropolitan Region induces high costs for transport and causes a considerable amount of traffic related air pollution. At present, Capetownians have to commute 16 km on average to their place of work (Gasson, 1995) what is equivalent to the distance people are travelling to their workplace in Los Angeles (however, with Los Angeles having five times more inhabitants). Dutkiewicz and de Villiers (1993) noted that the rates of air pollution as caused by the emissions of private and public transport, particularly during situations of inversion, range far above comparable values found elsewhere.

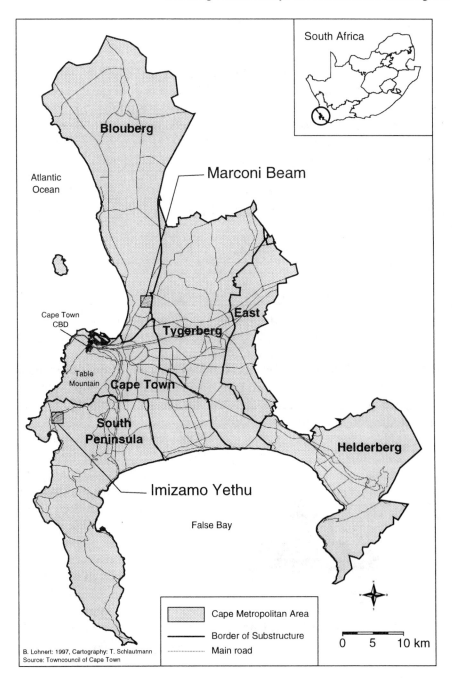

Figure 4.1   Location of the case studies

Two Case Studies: Marconi Beam and Imizamo Yethu

Two informal settlements in Cape Town are presented here as examples. The locations of the case studies are presented in Figure 4.1. Imizamo Yethu in Hout Bay and Marconi Beam in Milnerton both developed from illegal settlements and can be regarded as 'naturally' grown. A representative survey based on semi-structured interviews took place in late 1996 (Marconi Beam, n=185) and early 1997 (Imizamo Yethu, n=199) providing – among others – socio-economic baseline data as well as information about the migration history, ties to the rural areas of origin and the perception of environmental problems by the inhabitants. Additionally, qualitative interviews took place with all key stakeholders in the housing process.

Marconi Beam in Milnerton

The informal settlement of Marconi Beam is situated in the predominately white middle to upper middle class suburb of Milnerton and falls under the Blouberg substructure – see Figure 4.1. Milnerton is located about 8 km north-east of the city centre with a good transport infrastructure to town. Two big industrial areas – Montague Gardens Industrial Township which directly borders Marconi Beam and Metro Industrial Township – do not only provide jobs for Milnerton and thus for the inhabitants of Marconi Beam but also for other suburbs of Cape Town.

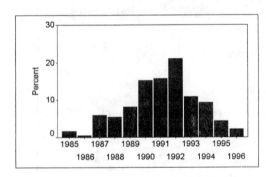

The – by then illegal – occupation of the Marconi Beam site started in the early eighties, with the majority of dwellers having arrived between 1990 and 1992 – see Figure 4.2. Massive protest against the settlement from white rate-payers of Milnerton resulted in the declaration of a 8.02 ha portion of the Marconi Beam site as 'transit area' under the *Prevention of Illegal Squatting Act* (1951) by the Municipality.

Figure 4.2   Arrival in Marconi Beam
Source: Own Survey.

In the following, a number of communal toilets and water-taps had been installed on the site. At the time of the survey (late 1996), approximately

5,700 people in 1,343 households lived in the transit area of Marconi Beam.

In 1994, after lengthy negotiations agreement was reached to resettle the Marconi Beam squatter camp further south to an area of 20 ha called *Joe Slovo Park*. The *Joe Slovo Park* site has been a quasi donation by the owner of the Marconi Beam site which is outlined for industrial development (Lohnert, 1999). At this moment, 1,000 government capital subsidised homes are in the process of being built in *Joe Slovo Park*. The so-called starter houses can be upgraded by the residents according to their financial resources. Most of them are built by contractors, some of them under organised self-help schemes. A non-repayable housing subsidy is granted according to household income. Households that earn less than R 800 per month qualify for the whole sum of 17,250 South African Rand which pays for site and service plus a basic top structure.

Imizamo Yethu in Hout Bay

Imizamo Yethu, being a Xhosa expression which stands for 'through our own struggle', is situated in Hout Bay, a suburb of Cape Town approximately 20 km south of the city centre, and falls under the South Peninsula substructure – see Figure 4.1. Hout Bay, a semi-rural residential area which is situated in a scenic valley has developed into a mainly white upper-income residential area during *Apartheid*. Inadequate provision for *non-white* people working and searching for work mainly in the harbour's fishing industry or as agricultural workers on the farms and domestics in the well-off households of Hout Bay led to the establishment of five informal settlements in the valley.

The settlements of *Disa River, David's Kraal* and *Blue Valley* were mainly inhabited by Afrikaans speaking *coloured* communities, while the much later established settlements *Princess Bush* and *Sea Products* were mainly inhabited by Xhosa speaking people. Especially *Princess Bush* and *Sea Products*, situated in the coastal dunes behind the prime recreation area of Hout Bay beach and on private land,

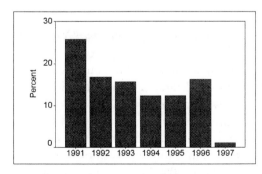

Figure 4.3 Arrival in Imizamo Yethu
Source: Own Survey.

evoked a strong opposition of the formal residents (Oelofse, 1996). After a fire in one of the informal settlements, and under pressure by formal residents, a resettlement of all informal areas to a Regional Forestry Council site onto successively serviced plots was agreed upon in 1991.

After that date the settlement experienced a major unplanned influx of new arrivals – see Figure 4.3. This increase resulted in half of the population presently living on unserviced sites of the steep slopes which are not suitable for low income housing (Lohnert, Oldfield and Parnell, 1998). Among the new arrivals were also some 500 to 700 non-South African citizens (McDonald, 1998). This immigration has led to severe conflicts between South Africans and foreigners. At the time of the survey in early 1997 approximately 4,500 people in 1,130 dwellings were living in Imizamo Yethu. At that time no housing project existed in Imizamo Yethu but negotiations on tenure rights with official bodies had started and contacts to facilitating organisations had been made. In contrast to the resettlement strategy in Marconi Beam, housing development in Imizamo Yethu will be heading towards *in situ* upgrading.

Re-creating Social Networks

As a major result of surveying the two settlement areas, it turned out that 85% of the migrants interviewed had come from rural areas of the Eastern Cape Province.[3] More than two thirds of the persons interviewed came directly to their present place of residence. This phenomenon and the analyses of the areas of origin of migrants in the respective settlements suggest a whole range of existing connections between the urban and the rural.

Origin of Urban Migrants and Working Mechanisms of Informal Networks

The survey indicated that 49% of the inhabitants of Marconi Beam who originate in the Eastern Cape (87% of the Marconi Beam sample) come from the Magisterial District Tsolo with another 21% having left the adjacent District Engcobo – see Figure 4.4. Asked for the main reason, why they came to Cape Town choosing Marconi Beam to put up a shack, nearly 50% stated that they wanted to be close to their work or near to job opportunities they had heard of from others. 10% said that they wanted to be near to their family and another 10% that they came here because they had been moved from their former place and they had noticed that 'their' people lived in Marconi Beam.

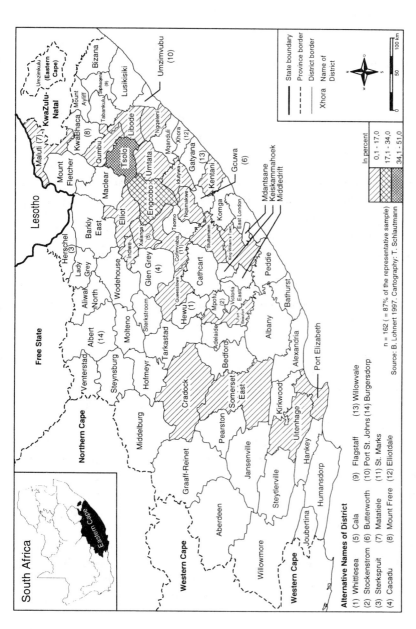

Figure 4.4 Origin of Marconi Beam dwellers

In a foreign environment, people tend to group with people that share their background and – very importantly – their language. Information is one of the most valuable assets shared within social networks. Migrants who visit their home area are multiplicators and very often constitute the only source of information for people intending to migrate. Therefore, the distinct majorities of people migrating from one or two magisterial districts to the naturally grown settlements are not astonishing as they resemble typical processes of chain migration. Information along the lines of informal social networks is not only shared on job opportunities but on all aspects that affect live in a strange environment. For migrants, being part of an information network is an important prerequisite for getting around in the urban environment.

As far as the re-creation of social networks and information flow is concerned, Imizamo Yethu shows a similar picture as Marconi Beam. 41% of the dwellers that originate in the Eastern Cape (83% of the Imizamo Yethu sample) came from the Magisterial District of Gatyana which is the former Willowvale District – see Figure 4.5. More than 50% named real or perceived job-opportunities as their major reason for migrating to Imizamo Yethu and 6% came to be near to their families.

The degree of integration into extended networks within the settlements basically depends on two variables, i.e., a shared background (family, place of origin, language, ethnicity) and the duration of stay in the community. People who came from the same region in their rural areas of origin had the best chances to be incorporated into such networks. Moreover, informal social networks are especially important for the elderly, women and women-headed households. In conclusion, the re-creation of social networks by rural-urban migrants reveals to be an important coping strategy to survive in the urban environment.

However, the strategy selected could imply negative impacts such as overcrowding, if people value their social setting higher than the environmental problems they are exposed to. The vast majority (70%) said that they intend to stay with their actual community. However, both settlements show extreme patterns of densification in terms of population and buildings leaving extremely limited space for solutions of *in situ* upgrading. For Marconi Beam where a resettlement strategy for the whole community is being implemented, the problem of social network disruption through urban upgrading processes does not apply. Planning processes in Imizamo Yethu however, where *in situ* upgrading seems the most likely solution, will have to struggle with the reluctance of inhabitants to move to other sites. Especially the informal part of the settlement is by far too overcrowded to leave any space for infrastructural facilities.

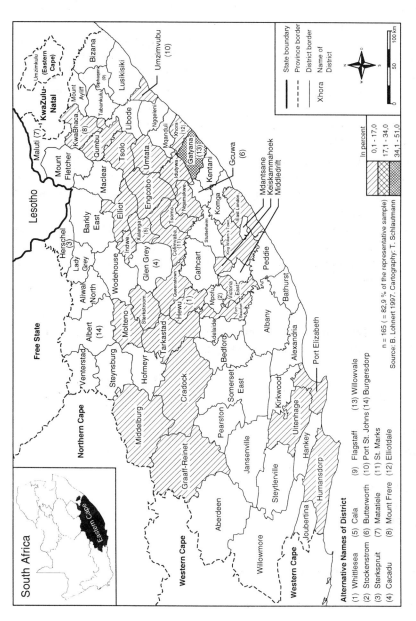

Figure 4.5 Origin of Imizamo Yethu dwellers

Informal Social Networks and Urban Planning

Extensive informal networks can also serve as indicators for the advanced consolidation of neighbourhoods. Social consolidation on the other hand is seen as a prerequisite for a successful self-help component in the upgrading process of physical living conditions. Social networks can contribute to a stronger identification of the individual with the place, thus enhancing the willingness to invest in the physical environment of urban informal settlements. The point has to be made here that minimal standards of land tenure security are required to enhance the acceptance of responsibility for the immediate environment (Barry, 1998; Mertins, Popp and Wehrmann, 1998).

Information on the social structure of informal settlements prove to be essential for sustainable planning concepts. This holds particularly true since resettlement and *in situ* upgrading is often planned 'from above' without adequate knowledge of existing social structures. However, from the viewpoint of participatory planning ('from below'), the recognition of the pattern of informational flows, degrees of consolidation, etc. is essential in triggering off successful negotiations on the proper formulation of upgrading policies.

Table 4.1   Main problems concerning living conditions: Marconi Beam and Imizamo Yethu inhabitants' perception*

Main Problems	Marconi B. (n = 185)	Imizamo Y. (n = 199)
Re-occuring fires	52 %	
Pollution of the neighbourhood, caused by garbage and faecal contamination	32 %	
Overcrowding	4 %	2%
Interior flooding of the dwellings	2 %	69%
Flooding of the area		11%
Others	10 %	18 %

* Data rounded to the nearest integer.

Source: Own Survey Data, 1996/97.

Environmental Problems: the Perception of the Migrants

As laid down before, inhabitants of informal settlements are exposed to a whole range of environmental risks. The outside view of planners and academics might, however, not necessarily match the inside view of people affected and will highly be dependent upon the actual setting.

As could be drawn from Table 4.1, the experiences with living conditions of the dwellers of the two settlements differ significantly when asked to specify and rank the main problems occurring.

Over the past years inhabitants of Marconi Beam have experienced a large number of fires. Especially in 1996 and 1997, fires caused deaths and the loss of many dwellings and material belongings. It is therefore not astonishing that the majority of interviewees in Marconi Beam named the threat of fires as their main problem – see Table 4.1. The use of kerosene for cooking and open fires for heating are regarded as the main reasons for the outbreaks of fires in that settlement. During the dry season, stormy winds which at times reach gale force level cause the spreading of sparks from open fires within the barely insulated shacks. The use of mainly inflammable building materials like cardboard and wood reinforces the danger of rapidly spreading fires. The increase of population and dwellings on the strictly demarcated area of the settlement led to a highly densified settlement structure – see Table 4.2.

Table 4.2   Population increase in Marconi Beam

Year	No. of dwellings	No. of persons	No. of persons per ha
1990	226	768	96
1991	500	2000	250
1993	834	2836	354
1994	1278	4856	605
1996	1343	5775	720

Sources: Urban Foundation (1993), *Marconi Beam, Interface Planning Case Study*, Cape Town; Saff, G. (1996), 'Claiming a Space in a Changing South Africa: The Squatters of Marconi Beam, Cape Town', *Annals of the Association of American Geographer*, vol. 2, pp. 235-55; Cape Town City Council, Department of Surveying and Geodetic Engineering, *Aerial Photographs 1996*, University of Cape Town; Own Survey Data 1996/97.

In some parts of the settlement, being apart from the major pathways, it is mostly impossible to walk between the shacks. Thus, the structure of the settlement is highly suitable to increase the possibility of single fires spreading from one shack to another.

Given second priority as a major problem of the environment, the pollution as caused by garbage and faecal contamination turns out to be closely related to the enormous population increase in Marconi Beam since 1991. The agreement between the Municipality and the inhabitants of Marconi Beam for a temporary stay in the 'transit area' in 1991 was followed by the building of a number of communal toilets and water-taps which met the immediate needs of the population size by then. In 1996, however, more than double the amount of people had to share these communal facilities. Additionally, ill-maintenance set a large number of the toilets out of order.

Flooding of houses and flooding of the entire area are the main problems mentioned by the inhabitants of Imizamo Yethu – see Table 4.1. Imizamo Yethu is built on the slopes of the Table Mountain Massif. While half of the population lives on serviced sites in the lower parts of the settlement, the other half that arrived after 1991 had to put up a shack on the steep slopes which are not suitable for low-income housing at all. The new arrivals moved further on to forest areas, thus, clearing the ground from trees and other vegetation. Continuing erosion processes induced by heavy rainfalls in winter were the results of this expansion. By now, the absorption capacity of the bare ground has deteriorated in a way that small streams cross the settlement during the rainy season. As in the case of Marconi Beam, unchecked population increase could be regarded as the major trigger of environmental problems experienced by the inhabitants of Imizamo Yethu.

Besides the above discussed main environmental problems, rural urban migrants are exposed to a whole range of other urban environmental threats. For this reason, the issue of environmental perception of 'urbanity' by rural urban migrants will be raised next.

Positive and Negative Aspects of Urban Life

To explore the urban experiences of migrants, people were asked to identify positive and negative aspects of living in town as opposed to living in rural areas. In Table 4.3, results are presented on the positive aspects perceived.

By most of the persons interviewed, 'economic opportunities' were given first priority in the mention of main positive aspects regarding urban life. This is in line with the reasons people stated as driving force to come to Cape Town. Considering some basic facts about the regions of origin, it is not astonishing that job opportunities are the main pull factors for rural-urban migrants. The combined effects of both economic negligence of the former *Homelands* during *Apartheid* and an immense population density, mainly caused by relocations and deportations from '*White Areas*', drove the socio-economic and environmental systems to the verge of collapse. In the rural areas of former Transkei and Ciskei the livelihood was and still is based on subsistence agriculture. However, the subsistence basis has been severely damaged due to food crop cultivation on unsuitable land and due to overgrazing. This resulted in ongoing erosion processes, thus, further destructing the subsistence basis.

Table 4.3   Positive aspects of living in town as opposed to living in rural areas (Marconi Beam and Imizamo Yethu)*

Economic opportunities	75%
(jobs, higher salary etc.)	
Physical infrastructure and services	13%
(water, transport, shops, toilets etc.)	
Social infrastructure	6%
(educational opportunities, medical support, etc.)	
No positive aspects	4%
Others	2%

* Data rounded to the nearest integer.

Source: Own Survey Data, 1996/97.

At present, 16% of the South African population live in the Eastern Cape Province with 24% of them bearly being able to satisfy their daily food needs of 2,500 kcal per head. In addition, Eastern Cape has the highest rate of unemployment (49%) to be found in the whole of the country (South African Labour and Development Research Unit, 1995).

On the background that 90 to 99% of the rural population in South Africa have no access to electricity and satisfactory sanitary infrastructure (Land and Agriculture Policy Centre, 1995, Development Bank of Southern Africa, 1996), it is not astonishing that informal urban dwellers who have at least access to communal taps, toilets, transport facilities and shops regard this part of urban infrastructure as one of the positive aspects in their urban life – see Table 4.3.

The negative aspects of urban life experienced by the interviewees, and presented in Table 4.4. show a typical picture of urban social problems. Especially in informal areas, crime and violence as well as alcohol and drug abuse are common features of every day life. A large proportion of the people interviewed had already been victims of criminal offences for themselves. Friday and Saturday are the crucial days with regard to violence. Most of the workers are paid weekly on Friday and spend most of their money in illegal liquor bars (*shebeens*) in the informal settlements over the weekend. Due to alcohol induced fights many deaths are recorded during the weekends in the informal settlements in Cape Town.

Table 4.4   Negative aspects of living in town as opposed to living in rural areas (Marconi Beam and Imizamo Yethu)*

Crime and violence	57%
Accidents	16%
Living in shacks	8%
Alcohol and drug abuse	7%
Bad hygienic conditions	7%
Others	7%

* Data rounded to the nearest integer.

Source: Own Survey data, 1996/97.

Conclusions

The Need for an Integrated Approach

It is often neglected that the enormous population increase of urban areas through migrants is – to a large extent – a function of the living conditions in the rural areas of origin. In this respect a holistic approach overcoming the urban bias is needed, tackling urban population increase and the housing question as a problem of urban as well as rural areas.

As data of the two case studies prove, most of the rural-urban migrants are in Cape Town because they don't see an economic and social perspective for themselves and their families in their home areas. Especially in the Eastern Cape which is still occupied with amalgamating three administrative structures (former Transkei, former Ciskei and the old Eastern Cape Province), economic development as well as housing policy do not yet have the same status of priority as it is given in urban areas.

In conclusion, there is a need for provinces to closely cooperate in future. This will particularly relate to the establishment of an institutional framework suitable for monitoring migration processes in order to react with adequate policy measures on both sides – the rural and the urban. Needless to say that the political will to interprovincial and interdepartmental co-operation is an absolute prerequisite. Without tackling the problems, which enforce rural-urban migration, housing policy in the great urban areas will become a (never ending) Sisyphean task.

The Role of Coping Strategies in Environmental Change Processes

With methodological regard to the assessment of human impact upon the urban as well as rural environment, it is suggested that coping strategies should be analysed in a broader context of societal structure and their long term effects should be monitored. As exemplified in the case of South African rural-urban migration, it is not the strategy as such that is to blame for environmental change processes but more so the underlying structural causes which create situations where coping measures have to be applied at all.

As for an integrated impact assessment, there is any reason to assume that rural-urban migrants are inclined to fall from one critical livelihood system into another simply by making use of one (or some) of the limited strategies they dispose of. From this, it is put forward here as an hypothesis on the social dimensions of environmental change that degrees of vulnerability will hardly change for the better. This is seen to hold true though there could be considerable variation in the type of risk, frame of time and spatial setting of processes constituting endangerment and threat.

In the context of global environmental change, the struggle for a decent life of the poor and vulnerable in the developing world must gain a first priority status on the agenda as it is very likely that a growing number of people will be forced to destroy environmental assets in order to survive.

Notes

[1]
The Group Areas Act (1950) was the legal basis for the establishment of separate residential areas for different race groups. The race groups were broadly defined as white, coloured, Indian and African. Although I do not accept these racial categories, it is inevitable to use them when reference is made to *Apartheid*.

[2]
Estimates from various experts interviewed 1997/98.

[3]
The Eastern Cape Province is the newly created administrative entity amalgamating the former 'independent' Homelands Transkei and Ciskei and the old Eastern Cape Province.

References

Association for the Promotion of the Western Cape's Economic Growth (1996), *Fact Sheet 2.2.*, WESGRO, Cape Town.

Barry, M. (1998), Proceedings of The International Conference on Land Tenure in The Developing World with a Focus on Southern Africa, January 27-29, 1998, Cape Town.

Barthlott, W. (1998), 'The Uneven Distribution of Global Biodiversity: A Challenge for Industrial and Developing Countries', in E. Ehlers and T. Krafft (eds), *German Global Change Research*, National Committee on Global Change Research, Bonn, p. 36.

Boucher, C. (1983), 'Western Cape Lowland Alien Vegetation', in *Proceedings of a Symposium on the Coastal Lowlands of the Western Cape*, University of the Western Cape, Cape Town.

Development Bank of Southern Africa (1996), *Baseline Data Study of the Eastern Cape*, DBSA, Pretoria.

Dutkiewicz, M. and de Vielliers, G. (1993), *Brown Haze in Cape Town, Air Pollution Update*, National Association for Clean Air, Stellenbosch.

Gasson, B. (1995), *Evaluating the Environmental Performance of Cities: The Case of Cape Town*, Paper presented at the IGU-Conference in Cape Town, August 21-25.

Land and Agriculture Policy Centre (1995), *Overview of the Transkei Sub-Region of the Eastern Cape*, LAPC, Port Elizabeth.

Lemon, A. (1991), *Homes Apart: South Africa's Segregated Cities*, Bloomington.

Lohnert, B (1998), 'Die Politische Ökologie der Land-Stadt Migration in Südafrika', *GAIA: Ecological Perspectives in Science, Humanities and Economics*, vol. 7 (4), pp. 265-70.

Lohnert, B (1999), 'From Shacks to Houses. Urban Development in Cape Town/South Africa: The Case of Marconi Beam', in A. Aguilar (ed) Proceedings of the IGU Commission Meeting on Urban Development and Urban Life, 1997, Mexico City.

Lohnert, B., Oldfield, S. and Parnell, S. (1998), 'Post-Apartheid Social Polarisations: The Creation of Sub-Urban Identities in Cape Town', *South African Geographical Journal*, vol. 80 (2), pp. 86-92.

Mathéy, K. (1992), *Beyond Self-Help Housing*, Mansell, München, London.

McDonald, D. (1998), *Left Out in the Cold? Housing and Immigration in the New South Africa*, Migration Policy Series No. 4, Cape Town.

Mertins, G., Popp, J. and Wehrmann, B. (1998), Bodenrecht und Bodenordnung in informellen großstädtischen Siedlungsgebieten von Entwicklungsländern, Beispiele aus Lateinamerika und Afrika, Fachbereich Geographie der Philipps-Universität Marburg.

Metropolitan Spatial Development Framework (1995), *A Guide for Spatial Development in the Cape Metropolitan Functional Region*, MSDF, Cape Town.

Meyer, W. B. and Turner II, B. L. (1995), 'The Earth Transformed: Trends, Trajectories, and Patterns' in R.J. Johnston, P.J. Taylor and M.J. Watts (eds), *Geographies of Global Change,*Blackwell, pp. 302-17.

Obudho, R. and Mhlanga, C.C. (1988), *Slums and Squatter Settlements in Sub-Saharan Africa, Towards a Planning Strategy*, New York.

Oelofse, C. (1996), 'The Integration of Three Disparate Communities: The Myths and Realities Facing Hout Bay, Cape Town', in R.J. Davies (ed) *Contemporary City Structuring, International Geographical Insights, Cape Town*, pp. 275-86.

Southern Africa Labour and Development Research Unit (1995), *Key Indicators of Poverty in South Africa*, SALDRU, Cape Town.

Takoli, C. (1998), 'Bridging the Divide: Rural-Urban Interactions and Livelihood Strategies', *iied, Gatekeeper Series*, No. 77.

United Nations Development Programme, World Bank and United Nations Centre for Human Settlements (1993), *Towards Environmental Strategies for Cities: Policy Considerations for Urban Environmental Management in Developing Countries*, Strategy Framework Paper, United Nations, New York,.

Watts, M. and Bohle, H.-G. (1993), 'Hunger, Famine and the Space of Vulnerability', *GeoJournal*, vol. 30 (2), pp. 117-25.

Westen, A.C.M. van (1995), *Unsettled: Low-Income Housing and Mobility in Bamako, Mali*, Utrecht.

5 Soil Mining and Societal Responses
The case of tobacco in eastern Miombo highlands

HELMUT GEIST

Introduction

What makes the essential part of this paper is, first, to provide some arguments that tobacco is suitable to constitute a 'key agricultural system' under the terms as specified by the International Geosphere-Biosphere Programme's (IGBP) core project 'Global Change and Terrestrial Ecosystems' (GCTE) – though not covered there – with particular reference to growing countries of the developing world (Tinker and Ingram, 1996). Secondly, tobacco's politicised environment is outlined in terms of tobacco industry and tobacco control oriented views on the crop ('symbolic' or 'political ecology'). Thirdly, data are provided on tobacco's potentiality of mining the soils ('real ecology'), and, fourthly, this is related to the situation of local tobacco farmers in two growing areas selected from the highland zone of Southeast Africa. By doing so, a rather unorthodox mix of methods, concepts and tools is applied which, however, is considered appropriate under the 'hybrid paradigms' created by social and natural science research on environmental change.[1]

Towards a Political Ecology of Tobacco

What makes Tobacco a Special Case?

Tobacco of the *N. tabacum* species is a 'major world crop' (International Tobacco Growers' Association, 1996) that is grown in more than 120 countries with developing nations presently accounting for more than 80% of the world production (Food and Agriculture Organization of the United Nations, 1998 a). Thus, it is the world's most widely cultivated non-food crop as compared to coffee (grown in 59 countries), tea (38), jute (25) or sisal (15). Though the share of tobacco in all arable land is only less than 1% on a global scale, the crop's impact as a 'driving force for economic development' (Reemtsma, 1995) is seen important particularly in producer countries of the developing world (ITGA, 1992; United Nations Conference on Trade and Development, 1995). This relates to direct and indirect employment as well as to the crop's contribution to foreign exchange and government revenues among other backward and forward linkages well into economy and society giving it the label of 'brown gold'. An estimated 30 to 33 million people directly owe their livelihood in whole or in part to the growing and processing of tobacco, and about 60 to 100 million if all tobacco-related industries and processes were included with nearly 90% of the employed found in the developing world (ITGA, 1997; Chaloupka and Warner, 2000). As for comparison, claimedly 'only about 1.7 million people are involved in the cultivation of maize or sugar cane' (ITGA, 1997).

From its biological properties, the tobacco plant can be grown up to $60°$ N with most of the produce, however, originating from semihumid to semiarid areas between $40°$ N and S. In these zones, a long enough dry season (allowing for harvesting and curing the crop), frost-free days and low-cost conditions of production such as cheap labour and open access to natural resources are given (Akehurst, 1981; Andreae, 1981; Fraser 1986 a). Between 1700 and 1800, almost the entire global output of the then already commercially grown crop originated from the New World (in particular the Brazil and Chesapeake colonies), while for the following centuries tobacco cultivation spread rapidly throughout the world because of its value both for 19th century colonialism and to many post-colonial countries in Africa, Asia and Latin America reacting upon the demands of the world market during 20th century (Goodman, 1993). While in 'northern' producer countries tobacco production decreased from 2.0 to 1.6 million tons in 1970 to 1997, it markedly increased from 2.6 to 6.7 million tons in 'southern' producer countries of the developing world with the total land under tobacco rising there from 2.6 to 4.1 million hectares during the time period considered (FAO, 1998 a).

Debating the Benefits from Tobacco Production

With the tremendous rise and structural (global) shift of tobacco production – as well as consumption – into the developing world, growing concerns have been expressed about both the health hazards involved in terms of a 'tobacco epidemic' and about 'equity and economic vulnerability', i.e., the growing countries' dependency upon a small group of transnational corporations purchasing the bulk of tobacco produce in the 'south'. Further concerns were raised that directly point to the environmental (un)sustainability of the crop in terms of an 'energy' or 'fuelwood crisis' (Muller, 1978; Goodland, Watson and Ledec, 1984; World Bank, 1984; Chapman and Wong, 1990). Summarizing these concerns, it was concluded that tobacco poses 'a particularly difficult dilemma for development', since its production generates both a range of employment, income, foreign exchange and other cash contributing effects, while 'the damage to public health and to the environment in the long term appears substantially to outweigh the benefits' (Goodland, Watson and Ledec, 1984). This sort of an overall assessment also underlies the most recent initiative of the so-called Bellagio Group.[2]

The negative externalities to be associated with the natural environments where tobacco is commonly grown fall under three broad categories, i.e., the usage of large amounts of wood (likely to result in deforestation), the considerable usage of biocides (pesticides, insecticides, fungicides, etc.), and soil nutrient depletion (with resulting land degradation if not mitigated). While soil depletion is considered here in more detail, the two other externalities will briefly be outlined as follows.

Wood usage and deforestation Though the tobacco industry in developing countries generally shows a lower consumption than the household sector and is a relatively small consumer of wood or forest products on a global scale, in certain cases – but even on a broader geographical scale – the industry tends to accentuate fuelwood shortages and deforestation. In fact, it has been found that often the use of fuelwood by the industry was highest in those developing countries where wood shortages were becoming severe (Fraser, 1986 a). Also it is pertinent to stress that a high proportion of the tobacco growing areas in developing nations are within parts of the world identified by the Food and Agriculture Organization of the United Nations as being wood deficit or in prospective wood deficit situations (FAO, 1998 b). Evidence from the regional level of growing areas suggests that the tobacco sector is the leading commercial consumer of wood and forest products accounting for up to 20% of all regional consumption (Fraser, 1986 b; Booth and Clarke, 1994; Siddiqui and Rajabu, 1996; Misana, Mung'ong'o and Mukamuri, 1996).

On the basis of energy sources used in the curing of tobacco, and by considering farmers' degrees of self-sufficiency in wood, it had been estimated that about 20 (stacked) cubic metres of wood were used to produce one ton of (cured) tobacco as an annual average in 1991 to 1995

(Geist, 1999 a). By translating wood consumption into the equivalent of natural woody biomass area needed and depleted, it has further been estimated that wood-use related deforestation of natural forests and woodlands by tobacco occurs in next to 70 growing countries of the developing world. Varying degrees of emerging environmental criticality can be related to tobacco's impact and were identified in half of the growing countries affected with most of them situated in middle east, south, and east Asia, in south and caribbean America and in southeastern Africa (Geist, 1999 b). While Fraser (1986 a) cites claims originating from the Global 2000 report that 2.5 million ha of forests were cut annually for tobacco curing (12% of total annual deforestation), recent data suggest that deforestation in 1991 to 1995 has been lowered to about 200,000 ha per year which is next to 2% of annual global net losses of forest cover or the equivalent of about 5% of total annual deforestation in the growing countries affected (Geist 1999 b). Tobacco industry commissioned reports have sought to repudiate previous acknowledgement of the problem by playing down the issue. An often repeated claim is that 'deforestation associated with tobacco curing cannot currently be considered a significant negative externality' (ITGA, 1995, 1996).

Pesticide usage and soil degradation Less publicised are the thousands of additives and pesticide residues in cigarettes that not only directly impact the lives of smokers, but exact a deep toll on the livelihood system of tobacco farming societies and their natural growing environment. With the global shift of farming into the developing world, a range of health, social and environmental problems are brought about that have to be quietly absorbed by the respective farming communities there since they lack the protection by environmental, pesticide-related and labour regulations as enforced by political struggles in the 'northern' countries.

Tobacco requires the application of considerably large amounts of artificial inputs such as pesticides. Chollat-Traquet (1996) summarizes that the 'use of complex compounds carries the possibility of crop contamination with inherent danger to those who smoke or chew the leaf, contamination of land and water-supply with danger to local communities, and occupational hazards for farmers and their families'. It is an important consideration especially in developing countries of the tropical zone where soils are characterized by their low nutrient content. In general, and to maintain soil fertility, the extraction of nutrients has to be balanced by suitable inputs of commercial fertiliser – but also pesticides in order to stabilise plant growth. Especially when tobacco is cultivated on land with minimal rotation, there is a tendency for soils to become exhausted and for crop pests to become endemic. An alternative to dependence on artificial inputs is to exhaust soil fertility in one or two years, then to clear new land for another plot, what in the past had been (and to some extent still is) done by deforestation (Goodland, Watson and Ledec, 1984; World Bank, 1984; Chapman, 1994).

Hypothesis With particular reference to 'southern countries', it was made explicit by Manshard and Mäckel (1995) that 'the growing of tobacco particularly in the tropical zone of the developing world is highly debated since it bears plentiful detrimental consequences and side-effects', namely, 'wood extraction', 'forest clearance' and considerable 'soil erosion'. Drawing here on Janzen's (1988) observation that tropical dry forest areas – where most of the production has shifted to – are 'the most endangered major tropical ecosystem', the hypothesis is put forward that tobacco farming not only is a major social and economic driver of land use and cover changes, but that there is any reason to further assume that tobacco farming areas constitute threatened environments beyond the scope of what has become accepted as the standard and detrimental impact of commercial farming upon natural environments (Kasperson, Kasperson and Turner, 1995). By relating to Odum's (1989) notion of endangerment, the environments where tobacco is commonly grown are seen to constitute 'endangered ecosystems' in that the physiological necessities of life such as mineral nutrients – among others – are destructively exploited for the market demands of the international cigarette industry.

Nicotine, Profits and Soil Depletion

Though assumedly being part of standard knowledge in production ecology, soil depletion by tobacco and its resulting consequences have hitherto been given a specific mention only by Goodland, Watson and Ledec (1984, p. 56) stating that

> most tropical soils are characterized by low nutrient content, particularly by deficiencies of phosphorus, often nitrogen, sometimes potassium. Tobacco production therefore usually depends on commercial fertilizers, the prices of which (especially nitrogen) are rising so sharply that they are increasingly out of reach of most developing country farmers (...) An alternative to dependence on commercial fertilizer is to exhaust soil fertility in one or two years, then to clear new land (often by deforestation) for another plot.

Table 5.1   Losses of major mineral soil elements as removed by tobacco and other crops

Crops with a standardized yield (harvest) of 1 ton/ha	Nitrogen (kg/ha)	Phosphorus (kg/ha)	Potassium (kg/ha)
Tobacco* (n=4)	50	14	105
Food crops (n=9)	10	3	16
... maize (n=2)	13	2	5
... rice (n=2)	11	2	12
... sorghum (n=2)	13	2	4
... finger millet (n=1)	16	10	65
... cassava (n=1)	1	0	1
... sweet potatoes (n=1)	4	1	6
Cash crops (n=20)	49	12	51
... coffee (n=2)	100	22	142
... oil palm (n=2)	212	46	251
... banana (n=3)	9	1	31
... sugar (n=2)	1	1	1
... rubber/latex (n=2)	8	3	6
... sisal (n=3)	17	3	41
... coconuts (n=1)	46	13	42
... cocoa (n=1)	27	14	20
... cotton (n=1)	34	11	9
... groundnuts (n=1)	45	11	30
... soya beans (n=1)	45	15	22
... tea (n=1)	45	9	21

* Virginia/flue-cured (2), cuban filler (1) and unspecified (1).

Source: Akehurst, B.C. (1981), *Tobacco*, Longman, London, p. 138; Tiffen, M. and Mortimore, M. (1990), *Theory and Practice in Plantation Agriculture, An Economic Review*, Overseas Development Institute, London, p. 100; Webster, C.C. and Wilson, P.N. (1980), *Agriculture in the Tropics*, Longman, London, p. 207, 297; Wrigley, G. (1969), *Tropical Agriculture: The Development of Production*, Praeger, New York, Washington, DC, p. 142.

The authors made explicit that 'tobacco depletes soil nutrients at a much higher rate than many other crops, thus rapidly decreasing the life of soil' (Goodland, Watson and Ledec, 1984).[3]

Quantifying the Uptake of Soil Nutrients

On the basis of a recent compilation of data from as many sources as available, a quantitative format is given to tobacco's uptake of soil nutrients. Though there will be much variation of data due to varying degrees of soil and climate specific conditions as well as agricultural practice (e.g., plant residues returned or not, leaching of N included or not), tobacco's mining of the soils could clearly be seen from Table 5.1.

It is found that the tobacco plant depletes major soil nutrients like nitrogen (N), phosphorus (P) and potassium (K) at higher rates than any other food crop and at mostly higher rates than any other cash crop grown. On average, the mineral uptake by tobacco amounts to 50 kg/ha for nitrogen, 14 kg/ha for phosphorus, and a uniquely high 105 kg/ha for potassium. When compared to the aggregated class of food crops such as maize, rice and sorghum, but also when compared to the aggregated class of cash crops such as sugar cane and rubber, their mineral uptakes range below that of tobacco. Only two cash crops have been identified that outstrip the nutrient absorption of tobacco, being specifically oil palm and coffee.

In summary, and as it is specified in Table 5.2, tobacco mines farming soils at considerably higher rates than most other crops, i.e., 80 to 85% higher than food crops and still 2 to 50% higher than other cash crops compared. Especially potassium is needed and absorbed in uniquely large amounts with tobacco outweighing cash and food crops by a factor up to 6.

Causes of Soil Depletion

The causes of tobacco's high mineral absorption could be seen in the design of the crop as a smoke product reaching out for high profitability ('brown gold'). Properties of plant biology will merge with agricultural practices such as 'topping' and '(de)suckering' in order to attain high levels of nicotine enrichment, high yields and, hence, high commercial profit.

Plant biology While, in general, favourable plant growth depends upon deficient items such as water (absorbed from the soil) and sunlight plus carbon dioxide (absorbed from the leaves), in the special case of tobacco, these items are mineral elements directly obtained from the soil. Among the major mineral elements ('macronutrients') usually required by tobacco in relatively large quantities, nitrogen (N), phosphorus (P) and potassium (or potash) (K) are the 'three most important nutrients' (Akehurst, 1981).

Table 5.2   Tobacco's relative potential of mining the soils

	Tobacco (n=4)	Cash Crops (n=20)	Food crops (n=9)
Losses of N			
... kg/ha	50	49	10
... % of tobacco		98	20
... factor*		<1	4
Losses of P			
... kg/ha	14	12	3
... % of tobacco		86	21
... factor*		<1	4
Losses of K			
... kg/ha	105	51	16
... % of tobacco		49	15
... factor*		1	6

* Factor by which tobacco outweighs other crops.

Source: Data as drawn from Table 5.1.

While the uptake of N is vital for growth, especially the green colour and nicotine content, P is essential for nutrition, especially the promotion of ripening, and K has a positive influence on leaf combustibility and the (bright) colour of flue-cured leaf (being the main ingredient of light American or Virginia Blend cigarettes). N and P are concentrated during early growth, while potash (K) is absorbed towards the end of the growing period (Akehurst, 1981).

Agricultural practice The commercial profitability of the crop will mainly be a result of the yield attained depending upon the amount, size, and weight of green leaf produced. Thus, before ripening and harvest – and besides artificial inputs such as fertilisation – the measures of 'topping' and '(de)suckering' are carried out. They constitute means of manipulating natural plant growth toward delivering higher yields of around 10 to 15% (Reisch, 1989).

'Topping' is the manual removal of inflorescenses and is done with most cigarette and cigar tobaccos (not with oriental and light tobaccos). It forces all nutrients not to go into seed but into leaf production instead, making the upper leaves grow longer, wider, thicker and darker in colour in order to have a higher nicotine content. Furthermore, it will stimulate root growth, slow down ripening rate and, thus, drain more nutrients from the soil (Akehurst, 1981; Acland, 1985).

However, as an undesired result, topping stimulates the growth of suckers, which, if not removed by hand at least once a week, will cause more harm than the flower would have done if left. Removal by hand will sometimes be supported by chemical means ('suckercides') such as *Antak* and *Tabamex*. Thus, '(de)suckering' means the manual removal of suckers and has to be carried out several times for as soon as they are removed yet more suckers are stimulated to grow (Schütt, 1972; Franke, Bruchholz, Fröhlich, Hain, Heynoldt, Husz and Pfeiffer, 1994).

The impact of suckers upon the yields achieved will be tremendous in merely emerging from the leaf axil, with the first 3 to 5 cm of emerging suckers being likely to cause the most damage. Commonly, they are not allowed to grow any longer than 7 to 13 cm in order not to seriously reduce yield (and profit). In the case of air-cured burley grown in Malawi, 'about 1% of yield is lost for every day suckers are left unremoved on the plant' (Tobacco Research Institute of Malawi, 1994). And, on fire-cured tobacco it is stated that a malawian 'farmer would lose an equivalent 100 kg of cured leaf per hectare', if 'suckers are allowed to grow to a length of 3 cm' (Mittawa, 1985).

'Sucking out the Heart of the Ground'

In summary, the measure of 'topping' induces a soil born nutrient drain into roots and leaves, while additional manipulation of natural plant growth is done in the form of multiple '(de)suckering' in order not to mitigate the desired nutrient flow into wide leaves with more body, thus, maintaining and reinforcing the flow of nutrients >from the soil. Since the measure of 'pruning' coffee shrubs and oil palm trees is similar but with larger amounts of woody biomass involved (Wrigley, 1969), this assumedly is one of the reasons for these crops to outweigh the already high nutrient uptake of tobacco.

As a consequence, tobacco growing if not mitigated will chiefly cause pronounced potash deficiencies especially in the early phase of the production cycle on newly cleared land. Thus, and in total, the crop has gained the title of a greedy and 'ruinous crop' (Madeley, 1986). As a matter of fact, evidence of soil mining could be traced from early colonial days up to present farming with reported incidences from next to all major growing areas of the developing world (Tobler and Ulbricht, 1942; Wilde, 1967; Silva, 1971; Abeysekera, 1985; Aliro, 1993; Goodman, 1993; Barickman, 1998).

Indications as such suggest that tobacco's unique depletion of macronutrients impedes natural soil nutrient balance thus reducing soil fertility at high rates, raises the risk for crop pests to become endemic, and establishes a twofold impetus either directed towards the increased usage of biocides (under the mode of rotational farming) or the renewed clearing of land and felling of trees (under the mode of shifting cultivation), besides – and thus accentuating – the impetus towards deforestation arising from the mere requirement of wood in curing.

Regional Setting and Research Design Used

Dry Forests Transformed

While considerably more and partly 'romanticized' attention is given to the transformation and destruction of humid tropical rainforests, less interest is devoted to the situation of dry forest and woodland ecosystems. This could partly be due to the fact that most of the wooded dryland areas in Asia and Latin America – such as *campos cerrados* or tree savannahs – have already been eliminated and transformed into agrarian (or partly even degraded and barren) land (Knapp, 1973). However, dry forest ecosystems cover about 7.7 million km^2 or the equivalent of 42% of tropical forest land and not only constitute the 'most endangered major tropical ecosystem' (Janzen, 1988), but more so does the majority of rural people in tropical countries depend on these ecosystems for their livelihoods (Campbell, 1996; FAO, 1997).

With the term 'drylands' being used in a broader sense, it was stated by FAO (1997) that these areas are inhabited by a large proportion of people who are among the world's poorest and that they 'are among the world's most fragile ecosystems and are made more so by periodic droughts and the risk of desertification' due to land degradation caused by climate and human activities (i.e., large-scale deforestation mainly for conversion to agricultural uses and overexploitation of forests and woodlands through fuelwood collection). Steinlin (1994) pointed out that in 1981 to 1990 tropical forest cover both in uplands and lowlands was lowest in dry forest zones (25 and 15% as compared to 44 and 29% in the humid parts), while deforestation progressed more in the dry lowland and upland forest zones (0.9 and 1.1% annually) as compared to the rain forest (0.6%) or total tropical forest zone (0.8%). Furthermore, population densities were highest in lowland and upland dry forest formations (106 and 70 inhabitants/km^2) as compared to the total tropical forest zone (52 inhabitants/km^2), with annual population growth in dry upland areas further ranking higher (3.2%) than in the total tropical (2.7%) or rain forest zones (2.5%).

The Case of Miombo Woodlands

The tropical dry forest zone of Africa holds the bulk of the world's land under dry forest and woodlands (i.e., 5.5 out of a total of 7.7 million km^2), while more than half of the african land falls under the world's largest and still more or less contiguous miombo zone of southeastern and central Africa (4.3 million km^2). Miombo is a vernacular word that has been adopted by ecologists to describe woodlands dominated by trees in the genera *Brachystegia*, *Julbernardia* and *Isoberlinia* with the total geo-ecosystem extending from Tanzania and Congo in the north, through Zambia, Malawi and eastern Angola, to Zimbabwe and Mozambique in the south. It features a hot, seasonally wet climatic zone the soils of which – mainly derived from acid crystalline bedrock – are predominantly infertile (Millington, Critchley, Douglas and Ryan, 1994; Desanker, Frost and Scholes, 1997).

Miombo woodlands have contracted to their present pattern of distribution since about 1,000 years and much of them are heavily modified to mostly secondary formations – with ecosystem transformation also encompassing the key wetland resources (*dambos*). Measured against degrees of 'human disturbance' and based upon data from the 1970s and 1980s, it was assumed that about 62% of the woodland cover had then still been 'undisturbed' (Hannah, Lohse, Hutchinson, Carr and Lankerani, 1994).

However, all factors so far contributing to the conservation of the woodlands are seen to have drastically changed during the most recent past (Campbell, 1996). On the present dynamics of miombo transformation it is stated that the driving factors are mainly 'macro-level phenomena' such as national policies related to land, agriculture, forestry, energy and population which have grown in strength and extent since about the last 20 to 30 years. Summarizing the historical perspectives on miombo utilisation,

Misana, Mung'ong'o and Mukamuri (1996) point out that woodland 'expansion that were characteristic of the pre-colonial and early colonial periods have now given way to continuous contraction in response to the ever expanding subsistence and commercial activities'. Rooted in colonial interventions, the agricultural economies based on tobacco and other export and colonial settler crops such as tea, coffee and cotton 'were increasingly drawn into the world market' with production 'intensification (including) ... both commerical and smallholder farms in order to meet market demands for export crops'. Specifically on tobacco, Misana, Mung'ong'o and Mukamuri (1996, p. 83) note that it

> became an important crop in the woodland areas of Tanzania, Malawi and Zimbabwe, and its increased production was encouraged through incentive schemes such as technical assistance and financial credit for mechanised operations, fertilisers and other equipment (...) Such intensification led to accelerated conversion of woodland areas to crops ... and increased demand for fuel for tobacco curing.

Drawing two case studies from this group of producers situated in the eastern part of the miombo zone, the countries' common features are that (i) they still have 25 to 48% of their respective country surface under miombo cover, (ii) they experience (above) average rates of national deforestation (0.6 to 1.6%), (iii) they produce a combined total of about 75% of the overall african tobacco output, and (iv) they have expanded their land under tobacco by about 30% in 1990 to 1995 (Geist, 1998 a, b). However, 'although it is generally agreed that miombo and associated woodlands are increasingly being cleared, the extent of such deforestation is unknown due to paucity of data' (Misana, Mung'ong'o and Mukamuri, 1996).

Criteria for the Selection of Study Areas

With one of the survey areas (Songea District) situated in southern Tanzania and the other (eastern Mangochi Dictrict) in southern Malawi, altogether six criteria are held suitable for a significant selection of two study areas as depicted in Figure 5.1.

First, the post-independence context of the countries' national political economy should be as distinct as possible. While Malawi is widely considered to practise a declared agro-capitalist mode of export-oriented world market production with tobacco accounting for 40 to 70% of annual foreign exchange revenues, Tanzania is known to have followed a distinct mode of self-reliant and socialist (*Ujamaa*) pattern of development with tobacco accounting for a considerably lesser percentage of foreign exchange revenues.

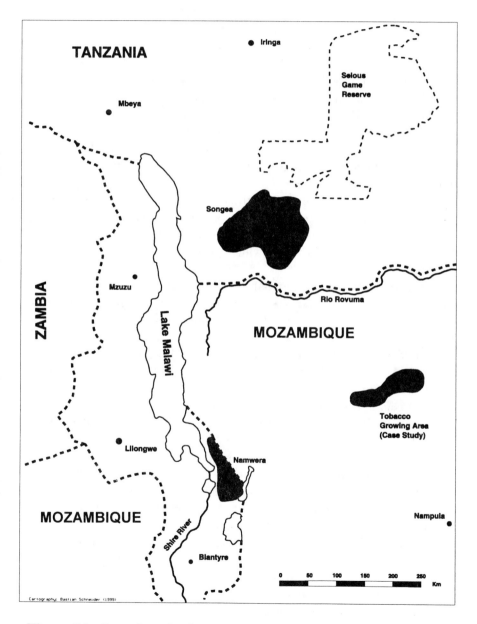

Figure 5.1 Location of tobacco growing areas in eastern miombo

Secondly, the regional ecological setting should be as similar as possible with both areas being situated in the highland zone of the Nubian (tectonic) plate east of the African Rift Valley at an altitude in the range of 800 to 1,200 metres. The areas are located at the northern and, respectively, southern fringe of a more or less contiguous miombo forest area that extends in between on the territory of northern Mozambique. The separating forest and agriculture zone has the approximate size of (former) East Germany and allows no interconnections between the study areas in terms of physical infrastructure, exchange of goods, services or people.

Thirdly, both areas are similar in that they constitute remote and marginal rural (*hinterland*) areas both bordering Mozambique. They had been provided with or linked to main national bituminous roads only as late as in mid-1980s. Both areas have experienced a considerable rural exodus as well as significant migrant labour movements since colonial times.

Fourthly, though separated in infrastructural terms, both areas not only bear comparable ecological features but are similarly shaped by the influence of a Yao-speaking moslem society with its origin in the Yao heartland of Nyassa Province which forms the northwestern part of Mozambique.

Fifthly, donor-driven regional economies are characteristic for both areas. From 1973 to mid-1980s, the Songea highlands fell under the multilateral donor activities of the Ruvuma Integrated Development Scheme (RIDEP) and since 1992 under the Dutch financed Songea Development Action (SODA). The Namwera highlands are part of the Liwonde Agricultural Development Programme (LWADD) which aims at integrated rural development under the aegis of the National Rural Development Programme (NRDP). Specifically, the Namwera Rural Development Programme (RDP) was initially financed by the African Development Bank and managed by the German Agency for Technical Co-operation (GTZ) from 1977 up to 1992 when all externally financed development came to a previous end.

Sixthly, in both areas agriculture is the backbone of the economy and tobacco forms a prominent factor of the agrarian setting. From Songea District, next to all of the Tanzanian fire-cured tobacco production originates constituting more than 10% of the variety's world production, and Namwera RDP area holds nearly half of the national flue-cured tobacco production of Malawi besides some nationally minor shares of air-cured burley and fire-cured tobacco.

Table 5.3 summarizes some of the highland areas' main features in terms of land and population figures as provided by official documents.

Most striking are the highly divergent population densities if related to the respective administrative areas (9 versus 87 inhabitants/km^2). However, if only farm population is related to land devoted to agriculture, the resulting values of agrarian density are more or less the same (around 140 inhabitants/km^2). It has to be noted, however, that in Namwera landless burley tenants and their families, which make up more than half of the tobacco farm population, are not represented by the official census or agricultural sample survey statistics as given here.

Table 5.3   Basic data of two miombo *hinterland* areas surveyed

	Songea highlands (Songea District)	Namwera highlands (Namera RDP)
Total surface (km^2)	36,300	1,900
... agricultural land (ha)	ca. 175,000	ca. 103,000
Total population	340,000 (1988)	166,000 (1989/90)
... farm population	253,000 (1988)	138,000 (1992/93)*
Population density	9 inh./km^2	87 inh./km^2
... agrarian density	145 inh./km^2	134 inh./km^2

* Not including estate tenant population.

Source: United Republic of Tanzania (1990), *Ruvuma: Regional Profile, Population Census 1988*, President's Office, Planning Commission, Bureau of Statistics, Dar es Salaam, pp. 206-22; Deutsche Gesellschaft für Technische Zusammenarbeit (1991), *LWADD Progress Review and Evaluation, Volume 1: Executive Summary and Main Report*, GTZ, Eschborn, p. 1 (annex); Government of Malawi (1996), *National Sample Survey of Agriculture 1992/93, Volume 1: Smallholder Household Composition Survey Report*, Ministry of Agriculture, National Statistical Office, Lilongwe, p. 193.

Research Design Used

From the next to spatial laboratory design thus established, i.e., testing the striking similarities of ecosystem endowment, socio-cultural background, physical isolation and impact of development programmes against the diverging contexts of national political economy, it was assumed that –

given tobacco's overall capacity of mining the soils – fairly reliable results on the political-economy impact upon managing the environment could be produced at the local level of tobacco farmers compared. However, it was neither assumed nor tested to what degree a socialist or capitalist economy impacts upon natural resources since, on the one hand, cash cropping is seen to set up a unilateral framework for specific capital/environment-relations and, on the other hand, the overall mechanisms of economic orders are known to ecologically bear more or less the same destructive features (Blaikie and Brookfield, 1987; Rauch, 1996).

Besides the use and evaluation of historical sources, census data, maps, satellite data, development agency reports, unstructured interviews and field observations, structured interviews with tobacco farmers identified to be responsible for land management decisions were central to the study. Interviews in both areas were done in July to September at the end of the growing season 1995/96, i.e., during the marketing period when the crop had already been harvested and cured. From a methodological point of view, sampling techniques were held different for smallholder and large-scale tobacco farmers due to their diverging shares in total farm population.

From the total of smallholder tobacco farmers reported to operate in the two growing areas, a random sample was taken ranging in both cases between 1 and 2% of all tobacco farmers registered. Originally, systematic random sampling of individual farmers was attained, either taken from the tobacco sales register of the Songea Agricultural Marketing Cooperative Union (SAMCU) or from LWADD tobacco files. If tobacco farmers thus sampled could personally not be identified in the field or at their home, another farmer operating next to his or her home or field was addressed while it was then agreed upon that preferably more female than male farmers should be included. In total, about 70% randomly selected farmers and 30% (gender) biased sampled farmers were included in the survey, i.e., 193 out of an estimated total of 18,600 farmers in Songea (1.0%) and 122 out of a registered total of 7,195 farmers in Namwera (1.7%). Results obtained are thus considered to be representative for the total of the smallholder tobacco farming population.

Large-scale tobacco operations (under the mode of estate farming) were only done in Namwera area. While a representative sample in the range of 7 to 70% of all estates was envisaged, out of a total of 369 estates altogether 113 of them were finally addressed (31.5%). In cases where the landlord, manager or *kapitao* agreed, additional interviewing was done with tobacco tenants operating on the farm. Thus, another altogether 137 tobacco farmers were included the respective sample percentage of which, however, is not known since their total size has never been covered by the national sample surveys of agriculture.

Be it a tobacco smallholder, tenant or large-scale farmer (landlord/ manager, hired manager or *kapitao*), interviewers hired from the respective Regional Departments of Agriculture and trained to use the questionnaire addressed the farmers in their local dialect. An interview was started if the farmer confirmed that he was the responsible land manager, i.e., making the day-to-day decisions in tobacco. Though standardized, the questionnaire, which consumed about 45 minutes of interview time, was designed to allow for specific management details of various types of tobacco grown (i.e., flue, burley and fire-cured). It was held shorter for tenants under the mode of burley contract farming.

Tobacco Farming Societies and their Environment

With regard to the local level processes studied, the issue of land is essential and three broad categories of persons holding land are distinguished (what also relates to the type of labour involved), i.e., (i) persons holding private and mostly large-scale land (and appropriating the surplus created by the direct work of others), (ii) persons under modes of common property regimes that seek to regulate resource use through specific management practices (and directly producing their means of livelihood including surplus), and (iii) persons holding no land at all (and simply struggling to reproduce themselves through wage labour for physical survival).

Social Structure, Land Rights and Landholdings

Reflecting highly divergent rural political economies, smallholder tobacco farmers of Songea turn out to have a rather egalitarian structure in that the medium size of landholdings is 4.5 ha, while Namweran farmers fall into several groups according to the status of their landhodlings, i.e., leasehold land for large-scale production on estates as a mean ranging between 24 and 880 ha, customary land for smallholder production ranging between 0.7 and 3.0 ha, and tenant land on estates normally in the order of 0.4 ha. It has to be noted that for most of the farming groups thus classified tobacco operations are part of overall and more diversified farming activities with tobacco holding on average between 10 and 30% of the total land.

Table 5.4 provides a breakdown of estate farmers in the Namwera highlands. The main difference in land management is that the smaller the estate, the more responsibility falls under the individual owner. While the larger estates are owned by foreign companies (such as Commonwealth Development Corporation), Greek landlords and members of the national ruling class (politicians, merchants, military and police staff) that have seized land since in the 1940s up to 1970s, most of the smaller estates are owned and run by graduated smallholders who reacted upon the immense land seizure by white and malawian landlords since the late 1980s. The mean number of tenants per farm ranges between 21 and 48, while the number of wage labourers declines with farm size from 728 (agrobusiness) to 64 (small estate). If all household members (but no tenants, wage labourers and people employed) are taken, the share of estate population in all tobacco farming population is only 3% which sharply contrasts the estates's share in all of Namwera's cultivable land being more than 50% (Geist, 1998 c).

Table 5.4   Typology of estate tobacco farms in Namwera

	Land (ha)	Tobacco Operation	Land manager
Agrobusiness (company)	200-2,500	Burley (mixed)*	hired professional
Large (individual)	200-1,200	Virginia (flue), Burley (mixed)*	mostly owner, partly *kapitao*
Medium	40-200	mostly Burley	owner
Small	<40	only Burley	owner

* Direct management and tenancy system.

Table 5.5 provides a breakdown of smallholder farmers in the Namwera highlands. While owner absentism in conjunction with hired management is characteristic for large estate holdings, smallholder farmers are solely responsible for their tobacco management decisions. Farm operations imply any transition from subsistence production with occasional tobacco market relation and wage labour to petty and extended commodity production. In contrast to estates, the part of wage labour is insignificant in favour of family labour recruited. While the growing of fire-cured tobacco under the smallholder mode of production has a long tradition in the area, the production of Burley has only been possible since early 1990s due to a crop act deregulation for the crop had previously only been allowed to grow on leasehold (estate) land by a small group of privileged farmers. If the sample is taken to be representative for the total farming population, smallholders account for about 43% of the tobacco society.

Table 5.5   Typology of smallholder tobacco farmers in Namwera

	Land (ha)	Tobacco Operation	Land manager
Large	1.8-12.1	Burley (direct), fire-cured	mostly full-time farmer
Medium	1.0-1.8	Burley (direct), fire-cured	mostly farmer
Small	<1.0	only Burley	part-time farmer

In the Songea highlands, the only difference among the rather homogenous group of tobacco peasants, who mostly make use of both family and hired labour, could be made up in terms of gender and household type. Hereby, it is meant that landholdings and tobacco plots of male-headed as well as polygamous farming households in general outweigh that of all other – and especially female-headed households – by factor 2.

Table 5.6   Typology of tenant tobacco farmers in Namwera

	Land (ha)	Tobacco Operation	Land manager
Large	0.4-2.1	only Burley	tenant farmer
Small	0.4	only Burley	tenant farmer and his family

Table 5.6 provides a breakdown of tobacco tenants in Namwera. The difference between the two groups is one of having access to land – or, at least, usage rights on customary land besides contract farming – or having not. The mass of tenants was found to be landless as opposed to the first

type of migrant smallholders (subsistence farmers) not taking up wage labour but entering contract farming instead. Though not represented in official statistics, tobacco tenants estimatedly account for the majority (54%) of the tobacco farming population in Namwera area (Geist, 1998 c).

Distinct Natural Growing Environments

From the scarce natural science information available on the growing areas, it could be stated that the less densely populated highlands of Songea has sufficient land resources and bears a forest cover still suitable for shifting cultivation around permanent settlements. Though satellite data identify the area as part of the 'wet' miombo zone, the area is seen to turn into a cultivation savannah being dominated by woody components like mangoes and Msuku ('Msuku woodland'). If widespread vegetation destruction is to be avoided in future, only slow rates of biomass exploitation were realistic (Millington, Critchley, Douglas and Ryan, 1994).

In the Namwera highlands, only the Mangochi and Namizimu Mountain ranges, now forest reserves, are covered by secondary growth of *Brachystegia* species. The plains, however, which were orginially covered by semi-natural vegetation consisting of *Brachystegia/Julbernardia/ Uapaca* regrowth with some woodland savannah, have been entirely cleared for for cultivation, and ground cover is almost non-existent (Venema, 1991). As a result of biomass assessment through satellite imagery, it was found that from mid-1970 to mid-1990 the forest cover has been reduced by 44% (in the total of Mangochi District), and that among all forest classes studied, in particular, the miombo species *Brachystegia* decreased by 85% – while thousands of hectares of newly logged forest areas could be observed even in protected zones (Government of Malawi, 1993).

Responding to Soil Mining: Inputs and Wood Consumption

Fertiliser and Biocides Used

Given tobacco's potentiality of mining the soils, in both of the growing areas surveyed the production of tobacco fully depends on commercial fertilizer, while the input of biocides to protect plant growth and mitigate soil exhaustion turns out to be highly different due to the unequal state of land degradation already occurred.

Songea Fertiliser in tobacco is used by next to all farmers (99%) in the form of 5 bags of predominantly sulphate of ammonia (S/A) as top dressing only. If the average amount is related to the mean size of tobacco plot cultivated, a total of next to 420 kg of fertiliser has been applied per hectare (with seedbed fertilisation not considered here). As can be seen from Table 5.7, the regionally adjusted fertiliser recommendations of the Agricultural Service exceed by far those for other crops (except for nitrogen), thus reflecting the crop's huge capacity of mining the soils.

Table 5.7   Fertiliser recommendations (kg/ha) in Songea District

Crop	Nitrogen	Phosphorus	Potassium
Tobacco	70	120	90
Maize	70	40	30
Beans		60	
Groundnuts		60	
Finger Millet	30	50	

Source: Turuka, F.M. (1995), *Price Reform and Fertiliser Use in Small-holder Agriculture in Tanzania*, Rural Development in Africa, Asia, and Latin America No. 51, Lit, Münster, Hamburg, p. 250.

Biocides, however, are only used by 3% of the farmers. Moreover, just half of these inputs are artificial (chemical) materials such as Novationsole, Thiodan and Aldrin – the latter of which is already phased out and legally prohibited in the developed world – , with the other half being organic inputs (in particular, Neem derived materials). What could be taken as a perfect indicator of sufficiently fertile grounds, no chemicals at all were applied on seedbeds (nurseries).

The main reason behind the low level of biocide (and, comparatively, also fertiliser) input is readily available land for shifting the fields including wooded areas to be cleared for new and fertile plots, what is also reflected in the region's overall low population density. As can be drawn from Table 5.8, the rate of land clearance for tobacco has considerably accelerated during the past 15 years with a tendency towards felling of trees on an annual basis. Correspondingly, it was confirmed by most farmers (88%) that there was a clearly observable decline in the number of trees during the period considered (1991-1996).

Table 5.8   Land clearance for tobacco in the Songea Highlands

	1 year	2-4 years	5-10 years	not at all
	Land clearance practised by farmers every ...			
1980-1990	18%	10%	14%	58%
1991-1996	45%	34%	7%	14%

Namwera Fertiliser is applied in the form of both basal and top dressing by all farmers and at considerably higher rates than in the Songea highlands (seedbed fertilisation not considered). Among the 11 (4) varieties used as basal (top) dressing, D-compound (CAN or Calcium Ammonium Nitrate) has been most widely used. The comparatively higher levels of input – see Table 5.9 – translate the fact that rotational cropping is carried out due to high overall population density and prevalent land scarcity (among smallholder farmers) with a tendency towards shortening or entirely eliminating fallow periods.

Table 5.9   Application of fertiliser on tobacco plots (kg/ha)

Type of farmers	Basal Dressing	Top Dressing	Total Dressing
Estates (Namwera)*	467	273	740
Smallholders (Namwera)	305	301	606
Smallholders (Songea)		417	417

* Tenants included.

The most striking difference, however, as compared to the rather fertile tobacco grounds of Songea is the high number and widespread usage of biocides applied on the fairly depleted Namweran soils, i.e., among all estate and 36% of the smallholder farmers. In particular, the intensive usage of biocides in nurseries could be taken as a clear indicator of no longer existing virgin soils to raise the seedlings. Among the most common biocides applied are Copper solutions, Azodrin, Thiram, Sevin and EDB with some of the biocides used (such as DDT) since long prohibited in 'northern' countries – see Table 5.10.

Table 5.10 Number of biocide varieties used in tobacco

Type of farmers	Basal Dressing	Top Dressing	Seedbed Dressing
Estates (Namwera)	19*	14	38*
Smallholders (Namwera)	2		12*
Smallholders (Songea)		3	

* Including DDT.

Wood Consumption for Curing Tobacco

Wood in farming is principally used as fuel for curing, but smaller amounts are also directly used in the form of poles and sticks in barn construction (storing). As in many rural areas of the developing world, wood is the most accessible and cost-effective source of construction material and of energy in the artificial curing of tobacco. Indigenous species such as Muyombo, Msuku and Muwanga are particularly favoured what is illustrated by the Songea tobacco farmers' preference for individual trees – see Table 5.11.

Table 5.11 Songea farmers' use of trees in curing and storing tobacco

Tree species	Curing (N=193)	Storing (N=183)
Muyombo (*Brachystegia floribunda*)	124	97
11 other miombo trees	5	8
Msuku (*Uapaca kirkiana*)	135	37
Muwanga (*Afromosia angolensis*)	10	103
Mtumbitumbi (*Cussonia angolensis*)	84	6
Mbuni/Mbula (*Parinari mabola*)	10	14
3 other indigenous trees	14	10
8 other indigenous (non-miombo) trees	25	41
2 exotic trees (*Cedrella, Gmelina arborea*)	18	4

As a forest product, wood is commonly taken from open accessible natural forests and woodlands and regarded as a 'free good' which means that no payment or contribution is made towards the cost of replacement. Only where shortages have developed, the market price is rising to a level where investment in plantation forests is becoming attractive. From this, the difference in the present natural endowment of the two growing areas is most obvious from Table 5.12. It illustrates that still existing common natural forests and woodlands are the main source of obtaining wood in the Songea highlands, while in Namwera area these ecosystems are depleted to an extent that most of the wood has to be obtained otherwise. While this relates only to smallholder farmers, it has to be noted that 75% of the Namweran tobacco estates, though required by covenant to have 10% of their land under tree cover, do not, so that most of the wood is either obtained through tractor transports from unspecified sources in Mozambique or otherwise brought/sold in.

Table 5.12 Sources of smallholder tobacco farmers' wood

Sources	Firewood for curing		Polewood for storing	
	Songea	Namwera	Songea	Namwera
Natural woodland and forests	94%		94%	
Private woodlot or communal plantation	16%	6%*	10%	12%
Forest Reserve, trader or taken from Mozambique	<1%	11%*	<1%	88%

* Only 21 fire-cured growers (multiple answers possible).

Curing means the transformation of green leaf tobacco into a pre-manufactured good mostly done on the farm. In general, tobacco when picked as a green leaf direct from the plant's stem must be cured to obtain the characteristic tobacco taste, aroma and colour and to preserve it for storage, transport and further processing. Among the four main methods of curing, natural curing makes use of the natural variations in temperature and humidity to dry up the leaves (air- and sun-curing), while artificial curing uses heat from energy sources such as coal, oil, gas and wood (flue- and fire-curing). In the case of flue-curing, heated air is passed through the harvested leaves by means of pipes (flues), while in the case of fire-curing wood smoke is introduced during the drying process to produce a dark, smoky-flavoured product.

Table 5.13 Farmwood consumption of artificially cured tobacco

N=Namwera S=Songea	One curing charge			All wood consumption	
	Mean number of poles	Mean stacked m^3	Mean no. of all curings	poles	stacked m^3
Flue estates (N) (n=35)	85	6.4	95.8	8,143	613
Fire-cured smallholders (N) (n=18)	12	0.8	9.8	118	8
Fire-cured smallholders (S) (n=144)	36	3.1	4.7	170	15

The process of curing puts heavy stress put upon the woody biomass resources in both of the growing areas. From Table 5.13 it can be seen that the average wood consumption of a tobacco farm for the sake of artificial curing is in the range of about 120 to 8,140 poles or the equivalent of 8 to 613 (stacked) cubic metres. Though fire-cured varieties were grown both in the Songea and Namwera highlands, the mean amount of poles consumed per one curing charge (and thus the total consumption) turns out to be lower in Namwera, since the Southern Division (Dark) Fire-Cured variety (SDDF) introduced by the Tobacco Service now consumes less wood than the 'Heavy' Northern Division Dark Fire-Cured tobacco (NDDF) used before – and still grown in Songea. Wood consumption is highest in flue-cured tobacco (Virginia) which as an american tobacco variety is exclusively grown under the mode of large-scale and capital intensive farming aimed at meeting the world market demands for light American and Virginia Blend cigarette tobaccos.

Conclusions

Though different in their national context of rural political economies, the case studies bear striking similarities in that they prove that money-making lies at the root of tobacco farming regardless of the impact upon natural environments. A fully disproportionate relation exists between ecosystems such as the dry forest or woodland zone of eastern miombo and the world market here exemplified in the form of international tobacco buyers, growers, processors and manufacturers intervening in rural *milieux* at the local level. With the on-going liberalisation of tobacco markets, smallholders in Tanzania seemedly realize their increased chances of income generation by entering market relations which had long been held unfavourable for them under the bureaucratic and inefficient mode of *Ujamaa*. Rent-seeking by European large farm lobbies had long been prevalent in other parts of Africa such as in Malawi. Aimed at reducing systematically the profitability of small farm cultivation through means such as unequal land distribution and contract farming, more of the smallholders start now to realize their increased income generation possibilities by entering tobacco market relations after national tobacco monopolies and market access have been deregulated. It is part of the contradictory relation between capital and environment that requirements for the sustainable management of natural resources are ranked last. Against the background of political efforts that increasingly shift from a demand-side perspective to supply-side measures of controlling the tobacco epidemic, as a matter of fair international relations, it will be of pivotal importance to open opportunities for a diversified and cash generating agricultural economy to smallholder farmers so as not to rob them income opportunities, but also to avoid that they undermine their production potentials and natural environments in the long-term.

Notes

[1] It was stated by Hard (1997) that natural scientists tend to use strictly natural science (or production ecology approaches) to tackle 'real' ecological phenomena as they perceive them, while they actually might obscure rather than explain ecological situations and problems. The reason could be seen in the fact that the actual state, change and variability of ecological phenomena not only depends on natural conditions but more so on the socially determined interpretations in terms of different 'symbolic' or 'political ecologies'. Modelling an ecology of the proposed broader type inevitably creates a 'hybrid paradigm' which had early been recognized by Enzensberger (1973).

[2] The Bellagio group constitutes an informal partnership which includes those UN and bilateral agencies, individual experts, research institutions, media, private sector groups, national agencies, foundations, and non-governmental organizations with particular interests in developing countries; see 'Bellagio Statement on Tobacco and Sustainable Development', *Can Med Assoc J*, vol. 153 (1), pp. 109-10; see also: <http:www.worldbank.org/html/extdr/hnp/hddflash/hcovp/ conf/conf002/htm>.

[3] Underlying the statement was Figure 8.3 providing the losses of N, P and K (in kg/ha) related to the harvest of one ton of tobacco, coffee, maize and cassava per hectare. However, the data given there could not be verified according to the original source as

specified, i.e., Wambeke, A. van (1975), 'Management Properties of Oxisols in Savanna Ecosystems', in E. Bornemisza and A. Alvarado (eds), *Soil Management in Tropical America*, North Carolina State University, Raleigh, pp. 364-71. Therefore, own data were compiled and produced as given in Tables 5.1 and 5.2 in the text.

References

Abeysekera, C. (1985), 'A Transnational in Peasant Agriculture: The Case of the Ceylon Tobacco Company', in C. Abeysekera (ed), *Capital and Peasant Production: Studies in the Continuity and Discontinuity of Agrarian Structure in Sri Lanka*, Social Scientists Association, Colombo, pp. 169-94.

Acland, J.D. (1985), *East African Crops: An Introduction to the Production of Field and Plantation Crops in Kenya, Tanzania and Uganda*, Longman, Harlow, Hong Kong.

Akehurst, B.C. (1981), *Tobacco*, Longman, London.

Aliro, K. (1993), *Uganda: Paying the Price of Growing Tobacco*, Monitor Publications, Kampala.

Andreae, B. (1981), *Farming, Development and Space: A World Agricultural Geography*, Gruyter, Berlin, New York.

Barickman, B.J. (1998), *A Bahian Counterpoint: Sugar, Tobacco, Cassava and Slavery in the Reconcavo, 1780-1860*, Stanford University Press, Stanford.

Blaikie, P. and Brookfield, H. (1987), *Land Degradation and Society*, Methuen, London.

Booth, A. and Clarke, J. (1994), 'Woodlands and Forests', in M. Chenje and P. Johnson (eds), *State of the Environment in Southern Africa*, Southern African Research & Documentation Centre in Collaboration with IUCN-The World Conservation Union and the Southern African Development Community Report, SARDC, IUCN, SADC, Harare, Masero, pp. 133-56.

Campbell, B. (ed) (1996), *The Miombo in Transition: Woodlands and Welfare in Africa*, Center for International Forestry Research, Bogor.

Chaloupka, F.J. and Warner, K.E. (2000), 'The Economics of Smoking', in J. Newhouse and A. Culyer (eds), *The Handbook of Health Economics*, University of Illinois at Chicago, University of Michigan (in press).

Chapman, S. (1994), 'Editorial: Tobacco and Deforestation in the Developing World', *Tobacco Control*, vol. 3 (3), pp. 191-93.

Chapman, S. and Wong, W.L. (1990), *Tobacco Control in the Third World: A Resource Atlas*, International Organization of Consumers Unions, Penang.

Chollat-Traquet, C. (1996), *Evaluating Tobacco Control Activities: Experiences and Guiding Principles*, World Health Organization of the United Nations, Geneva.

Desanker, P.V., Frost, P.G.H., Justice, C. and Scholes, R.J. (1997), *The Miombo Network, Framework for a Terrestrial Transect Study of Land-Use and Land-Cover Change in the Miombo Ecosystems of Central Africa: Conclusions of the Miombo Network Workshop, Zomba, Malawi, December 1995*, IGBP Report No.41, IGBP, ICSU, Stockholm.

Enzensberger, H.M. (1973), 'Zur Kritik der politischen Ökologie', *Kursbuch*, vol. 33, pp. 1-42.

Food and Agriculture Organization of the United Nations (1997), *State of the World's Forests 1997*, FAO, Rome.

Food and Agriculture Organization of the United Nations (1998 a), *Internet Data of Production of Primary Crops*, FAO, Rome, <apps.fao.org/lim500/nph-wrap.pl? Production.Crops.Primary&Domain=SUA>.

Food and Agriculture Organization of the United Nations (1998 b), *FAO Statement on Multisectoral Collaboration on Tobacco or Health for the ECOSOC Substantive Session*, FAO, Rome.

Franke, G., Bruchholz, H., Fröhlich, G., Hain, W., Heynoldt, H.-J., Husz, W., Pfeiffer, A. (1994), *Spezieller Pflanzenbau*, Nutzpflanzen der Tropen und Subtropen No. 3, Ulmer, Stuttgart.

Fraser, A.I. (1986 a), *The Use of Wood by the Tobacco Industry and the Ecological Implications*, International Forest Science Consultancy, Edinburgh.

Fraser, A.I. (1986 b), *The Use of Wood by the Tobacco Industry in Argentina*, International Forest Science Consultancy, Edinburgh.

Geist, H. (1998 a), 'Tropenwaldzerstörung durch Tabak: Eine These erörtert am Beispiel afrikanischer Miombowälder', *Geographische Rundschau*, vol. 50 (3), pp. 283-90.

Geist, H. (1998 b), 'How Tobacco Farming Contributes to Tropical Deforestation', in I. Abedian, R. van der Merwe, N. Wilkins and P. Jha (eds), *Economics of Tobacco Control: Towards an Optimal Policy Mix*, Applied Fiscal Research Center, University of Cape Town, pp. 232-44.

Geist, H. (1998 c), 'Das Bergland von Namwera: Eine Fallstudie über Landdegradierung, Gemeinheitsteilung und braunes Gold', *GAIA-Ecological Perspectives in Science, Humanities and Economics*, vol. 7 (4), pp. 255-64.

Geist, H. (1999 a), 'Transforming the Fringe: Tobacco-Related Wood Usage and its Environmental Implications', in F. Delgado-Cravidao, H. Jussila and R. Majoral (eds), *Consequences of Globalization and Deregulation on Marginal and Critical Region Economic Systems*, Ashgate, Aldershot (in press).

Geist, H. (1999 b), 'Global Assessment of Deforestation Related to Tobacco Farming', *Tobacco Control*, vol. 8 (1).

Goodland, R.J.A., Watson, C. and Ledec, G. (1984), *Environmental Management in Tropical Agriculture*, Westview Press, Boulder.

Goodman, J. (1993), *Tobacco in History: The Cultures of Dependence*, Routledge, London, New York.

Government of Malawi (1993), *Forest Resources Mapping and Biomass Assessment for Malawi*, Satelitbild, Ministry of Forestry and Natural Resources, Kinema, Lilongwe.

Hannah, L., Lohse, D., Hutchinson, C., Carr, J.L. and Lankerani, A. (1994), 'A Preliminary Inventory of Human Disturbance of World Ecosystems', *Ambio*, vol. 23 (4/5), pp. 246-50.

Hard, G. (1997), 'Was ist Stadtökologie? Argumente für eine Erweiterung des Aufmerksamkeitshorizonts ökologischer Forschung', *Erdkunde*, vol. 51, pp. 100-13.

International Tobacco Growers' Association (1992), *The Economic Significance of Tobacco Growing in Central and Southern Africa*, Bardwell's, Harare.

International Tobacco Growers' Asscociation (1996), *Deforestation and the Use of Wood for Curing Tobacco*, Tobacco Growers-Issues Papers No. 5, ITGA, East Grinstead.

International Tobacco Growers' Asscociation (1995), *Tobacco and the Environment*, Tobacco Briefing, ITGA, East Grinstead.

International Tobacco Growers' Asscociation (1997), *The Use of Woodfuel for Curing Tobacco*, Tobacco Growers-Issues Papers No. 11, ITGA, East Grinstead.

Janzen, D.H. (1988), 'Tropical Dry Forests: The Most Endangered Major Tropical Ecosystem', in E.O. Wilson (ed), *Biodiversity*, National Academy Press, Washington, D.C., pp. 130-37.

Kasperson, J.X., Kapserson, R.E. and Turner, B.L.II (eds), *Regions at Risk: Comparisons of Threatened Environments*, United Nations University Press, Tokyo, New York, Paris.

Knapp, R. (1973), Die Vegetation von Afrika unter Berücksichtigung von Umwelt, Entwicklung, Wirtschaft, Agrar- und Forstgeographie, Vegetationsmonographien der einzelnen Großräume No. 3, Fischer, Stuttgart.

Madeley, J. (1986), 'Tobacco: A Ruinous Crop', *The Ecologist*, vol. 16 (2/3), pp. 124-29.

Manshard, W. and Mäckel, R. (1995), *Umwelt und Entwicklung in den Tropen: Naturpotential und Landnutzung*, Wissenschaftliche Buchgesellschaft, Darmstadt.

Millington, A.C., Critchley, R.W., Douglas, T.D. and Ryan, P. (1994), *Estimating Woody Biomass in Sub-saharan Africa*, World Bank, Washington, DC.

Misana, S., Mung'ong'o, C. and Mukamuri, B. (1996), 'Miombo Woodlands in the Wider Context: Macro-Economic and Inter-Sectoral Influences', in B. Campbell (ed), *The Miombo in Transition: Woodlands and Welfare in Africa*, Center for International Forestry Research, Bogor, pp. 73-99.

Mittawa, G.I. (1985), *Tobacco Management Handbook: Notes for Field Extension Staff*, Ministry of Agriculture, Extension Aids Branch, Lilongwe.

Muller, M. (1978), *Tobacco and the Third World: Tomorrow's Epidemic ? A War on Want Investigation into the Production, Promotion, and Use of Tobacco in the Developing Countries*, War on Want, London.

Odum, E.P. (1989), *Ecology and our Endangered Life-Support Systems*, Sinauer Associates, Sunderland.

Rauch, T. (1996), *Ländliche Regionalentwicklung im Spannungsfeld zwischen Weltmarkt, Staatsmacht und kleinbäuerlichen Strategien*, Sozialwissenschaftliche Studien zu internationalen Problemen No. 202, Verlag für Entwicklungspolitik, Saarbrücken.

Reemtsma (1995), *Tobacco: Driving Force for Economic Development*, Reemtsma Cigarettenfabriken GmbH, Hamburg.

Reisch, W. (1989), 'Tabak', in S. Rehm (ed), *Spezieller Pflanzenbau in den Tropen und Subtropen*, Handbuch der Landwirtschaft und Ernährung in den Tropen und Subtropen No.4, Ulmer, Stuttgart, pp. 459-70.

Schütt, P. (1972), *Weltwirtschaftspflanzen: Herkunft, Anbauverhältnisse, Biologie und Verwendung der wichtigsten landwirtschaftlichen Nutzpflanzen*, Parey, Berlin, Hamburg.

Siddiqui, K.M. and Rajabu, H. (1996), 'Energy Efficiency in Current Tobacco-Curing Practice in Tanzania and its Consequences', *Energy*, vol. 21 (2), pp. 141-45.

Silva, C.N. (1971), *Das Tabakanbaugebiet des Reconcavo von Bahia/Brasilien*, Krause, Umkirch.

Steinlin, H. (1994), 'The Decline of Tropical Forests', *Quarterly Journal of International Agriculture*, vol. 33, pp. 128-37.

Tinker, P.B. and Ingram, J.S.I. (1996), 'The Work of Focus 3', in B. Walker and W. Steffen (eds), *Global Change and Terrestrial Ecosystems*, IGBP Book Series, Cambridge University Press, pp. 207-28.

Tobler, F. and Ulbricht, H. (1942), *Koloniale Nutzpflanzen: Ein Lehr- und Nachschlagebuch*, Hirzel, Leipzig.

Tobacco Research Institute of Malawi (1994), *Malawi Burley Tobacco Handbook*, TRIM, Lilongwe.

United Nations Conference on Trade and Development (1995), *Economic Role of Tobacco Production and Exports in Countries Depending on Tobacco as a Major Source of Income*, Report No. 51627, UNCTAD, Geneva.

Venema, J.H. (1991), *Land Resources Appraisal of Liwonde Agricultural Development Division*, LREP Report & Field Document No. 23, Government of Malawi, Ministry of

Agriculture and Livestock Development, Land Husbandry Branch, UNDP, FAO, Lilongwe.

Wilde, J.C. de (1967), *Experiences with Agricultural Development in Tropical Africa*, The Case Studies No. 2, Johns Hopkins Press, Baltimore.

World Bank (1984), *World Bank Tobacco Financing: The Environmental/Health Case, Background for Policy Formulation*, Office of Environmental and Scientific Affairs Projects Policy Department Paper No. W0020/0087W/C2402, World Bank, Washington, D.C.

Wrigley, G. (1969), *Tropical Agriculture: The Development of Production*, Praeger, New York, Washington, DC.

6 Changes of Land Use and Institutions in Natural Resource Management
The case of the Tanzanian Maasailand

SVEN SCHADE

Introduction

Land use change is more than a simple change of agricultural techniques. It provokes a change in the relative value of natural resources in the production process. According to North (1990), this is the starting point for a reform of institutional arrangements. Global environmental change and a global economic system are catalysing local land use changes and leading to new local institutional arrangements to regulate resource use. This will have feedback on the resilience of the ecosystem (Holling, 1973) as much as on the economic and social system.

This contribution presents how rapid land use change in the Northern Tanzanian drylands from Maasai's transhumant pastoralism to the production of seed-beans for export not only altered the availability of grazing resources but also the rules and the rule making bodies that regulate the use of resources.

Environmental and Social Setting of Naberera

Today the village area of Naberera in Tanzania is one of the world's leading centres of seed-bean production serving about 25% of the global market.[1] Only 20 years ago, it was a settlement area of transhumant pastoralist Maasai around important all-year watersources somewhere in the Maasaisteppe. The village is situated at 37° East and 4°12' South of the equator with the communal area covering about 1,100 km² and having a total population of 4,065 inhabitants in 1988 (United Republic of Tanzania, 1991). But population figures are unreliable due to the seasonal migration of

149

both pastoralists and others coming from neighbouring districts and involved in the production of charcoal.

Naberera receives a total of 430 mm rainfall annually in two periods, i.e., a short rainy-season in November/December and long rains from February to May. Rainfall is highly variable in time and space. During the period of field research in 1996/97, the total of the region was hit by a severe drought. Short rains completely failed and long rains began only on April 10 lasting until May 20. During the months of drought, the survival of Maasai herds was secured by an area of 4 km² that received showers. Several thousand head of cattle assembled there to benefit from sprouting herbs and grasses and the Naberera wells nearby still offering water.

The region's vegetation is dominated by 'open Acacia-Commiphora savannah woodlands' and 'open grassland with scattered trees' (National Environment Management Council, 1993). Some areas, specifically around isolated, steep hills rising from the plains at 1,200 m altitude up to 2,200 m, receive considerably more rainfall. These areas support dense woodlands that are an important local source of building materials and medicines.

Most of the woodlands in Naberera are infested by Tsetse flies (*Glossina swyntonneri*). These are vectors for cattle trypanosomiasis, so large areas can be used for grazing only in times of need when adequate fodder resources in non-infested areas are no longer available.

Maasai Traditional Land Use

Maasai took over the area from pastoralist Parakuyu during the Iloikop wars in the 1840s to 1850s. Control over the wells in Naberera meant control over the grazing resources in much of the southern Maasaisteppe. Since the Iloikop wars the land was used for the production of milk, blood and meat (in decreasing order of importance for nutrition) and of cattle for trade and social purposes. In times of food scarcity, animals were exchanged for cereals with settled farmers around Mount Meru and Mount Kilimandjaro (Waller, 1985). These social links were crucial to survive the effects of the devastating rinderpest epidemics and the droughts in the 1890s. During that time Maasai took up cultivation in better suited areas in order to restore the herds.

All through the colonial period and up to the recent past Maasai practised transhumant pastoralism. In areas bordering farming communities, Maasai mixed with groups of other ethnic origin and began smallholder farming (agropastoralism). In transhumant pastoralism, at the beginning of the short rains circumcised men (*Ilmurran* or 'warriors') and older boys move the herds to areas about 20 to 50 kilometres away that offer fresh grazing and sufficient water in rainponds. When these have dried up, the herds are brought back to the homestead of the household (*engkang*). More members of the household will move at the beginning of the long rains. Then, only elder women and men were left behind with a number of cattle, some women, a few herderboys and girls. When the seasonal sources of water have dried up, the whole household will reunite in the old *engkang*, or a new *engkang* is built somewhere else near a permanent source of water – see Figure 6.1. The local prevalence of game as vectors of diseases (such as malignant catarrh) and of ticks results in considerable variations in the migrational pattern (Potkanski, 1998).

Droughts are frequent, but rarely do they affect all of the Maasailand. Surviving a severe drought is only possible if the household can gain access to water and grazing resources in areas further away. Ndagala (1992) and Potkanski (1998) give excellent overviews on the mechanism of how Maasai secure access to resources by social organisation in age-sets, clans, sections, through marriage and 'cattle-friendships'.

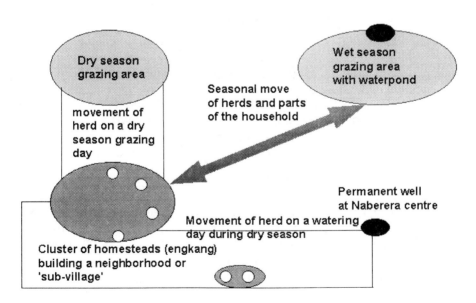

Figure 6.1 Traditional transhumant pastoralism in Naberera

Social Organisation and Access to Resources

Non-spatial social organisations include age-sets, clans and cattle-friendships, while spatial forms include the so-called section,[2] locality (*enkutot*) and neighborhood (*engutoto*) – see Figure 6.2.

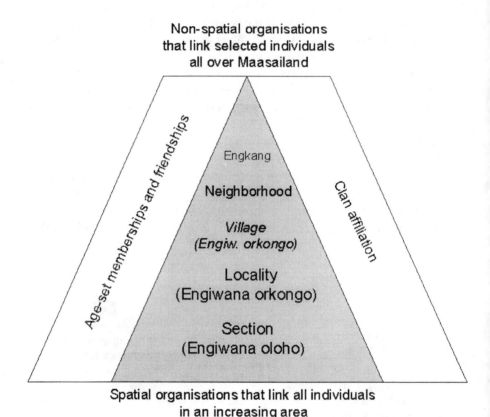

Figure 6.2 Social organisation of Maasai society

Non-spatial organisations As with many other pastoral societies in East-Africa, Maasai men are grouped into age-sets according to the time of circumcision. Periods for circumcisions started in 1935, 1955 (for the *Seuri* age-set), 1971 (for the *Makaa*), 1977 (for the Landis) and 1998 (name not yet known). Circumcision periods last for 5 to 10 years (Ndagala, 1992). The youngest boys to be circumcised are around 13 years old. Following circumcision boys become members of the *Ilmurran* age-class of which they are members for 16 to 23 years depending on the time of their circumcision and the time of opening of a new age-set. Only by finishing the *Ilmurran* period, young men gain full social maturity. Belonging to the same age-set means special obligations to age-mates such as offering housing when on move and others.

According to Rigby (1988), livestock is a means of production and not the product of Maasai pastoralism. The clan owns the cattle (Potkanski, 1998), although the male household head has the right to dispose of or slaughter the animal if necessary. At the same time, he has to allocate sufficient heads of livestock to his wives so that they can distribute milk to children, men and women. The clan's property of cattle becomes most obvious when a clan member is in need such as after droughts or outbreaks of herd diseases. Then, he might approach a clan meeting for donations of livestock. This can only be refused if the loss of the herd was the result of unacceptable individual behaviour (Potkanski, 1998). The clan members are also approached if a fine for murder has to be paid to another clan, commonly being 29 or 49 heads of cattle.

'Cattle-friendships' constitute an individual network of personal contacts between men. These are not restricted to clan members or age-mates. By a gift of livestock a 'brother-like' relation is established that insures support in times of need.

All these non-spatial social organisations secure access to spatially variable resources. They facilitate mobility and increase the chance of herd survival in times of scarcity. Membership in non-spatial organisations establishes entitlements to resources. The question, however, arises whether these institutions are also in charge of the management of spatially distinct resources.

Spatial organisations   The largest spatial unit of social organisation in Maasai society is the section. Out of about ten sections, Kisongo-Maasai is the largest and covers most of the Tanzanian Maasailand. Membership to a section is defined by the acceptance of a spiritual leader (*Olaibon*). During the past, conflict over resources was abundant between the sections (Waller, 1985). As the spiritual leader commands the forces of the *Ilmurran* (warrior age-class), distinct but overlapping settlement areas for different sections have developed.

The spiritual leaders have representatives (*Olaigwanani*) in each locality (*enkutot*). The locality is a spatial unit that covers several settlement centres. It had been suggested by Ole Ngulay (undated) to look upon locality as an ecological unit because all types of ecosystems essential for pastoral production can be found within a single locality. A special committee of elders elects the local representatives of the *Olaibon* from the members of the *Ilmurran* age-class.

Meetings of the socially mature Maasai men take place at locality, village and neighborhood level. Senior *Murran* participate in these meetings although they traditionally won't own a herd at that time. The locality of Naperera (Inyuat-e-Maa, 1991) covers Naberera and two villages nearby totalling an area of more than 3,000 km². Although there were meetings to discuss local issues on 'sub-locality' level in the past, the observation that more issues are nowadays discussed by the elders living in 'modern' Naberera village (1,000 km²) can be attributed to the 'Ujamaa-village policy' of the Tanzanian government (Ndagala, 1992).

State administration   The hierarchical order of statal administration comprises the national, regional, district, ward and village level. In 1996/97, Tanzania was at the height of reforming a confuse land law situation. From the tremendous amount of literature on land law reform, the reports by the Land Commission (1994) and Shivji (1995) could be considered to the most significant. The main features of current land law are summarized there as follows:

- all land is vested in the president;
- the state grants 'rights of occupancy';
- registered villages can set up village land use plans and have the right to distribute land to villagers and outsiders;
- if larger parcels of land (i.e., more than 50 acres) are allocated, the district, region or ministry have to approve the allocation according to the size of the land under consideration; and
- the village council has to approve any allocation to outsiders.

The heads of the regional and district administration are Commissioners appointed by the President. Heads of regional and district departments are regional and district officers. Elected councils exist on each level.

The village assembly is a meeting of all adult villagers. It elects the village chairman and the village council. At least 5 of the 25 members of the village council have to be women. A village secretary is the right hand of the village chairman. Village officers for development, livestock, game, agriculture and health are members of the respective committees of Naberera village council.

Parallel to the state administration, a structure of the former Party of National Unity (Chama cha mapinduzi, CCM) used to be in operation from the national level down to level of neighborhoods (Shivji, 1995). Although this structure no longer exists, many local elders officially still introduce themselves as 'ten-cell leader' which used to be the party spokesman for a total of ten households. Loiske (1995) and Lindberg (1996) report that in the past many decisions on resource use were made within the party structure.

Present Systems of Land Use

At present, most of Maasai households cultivate maize for home consumption. As beans are not appreciated as food, they are cultivated to a limited extent and mainly as cash crop. The majority of households (i.e., husband, wives and children) cultivate a plot of 10 to 30 acres. During the 1990s, prices of livestock fell because of a decline in the demand for meat in urban centres due to the policy measures of structural adjustment resulting in tremendous real income losses on the side of urban based state officers and employees (Schade and Ibrahim, 1999).

Charcoal production affects an increasing part of the savannah woodlands. In the 1960s, charcoal destined for the town of Arusha was produced 40 km south of town near the settlement Oljoro No. 5, while the present centres of production are situated near Naberera which is 120 km south of Arusha. Two types of charcoal producers can be identified:

- married men with children too young to provide labour on small family farms; due to cash requirements for the provision of food, school fees and medicine, married men migrate to charcoal production areas and return before the start of the farming season; and
- small families searching for a plot to farm; the families are dependent on charcoal as an additional source of income since dryland farming often fails; though they would prefer to settle permanently, they are forced to move on as soon as wood resources become scare and no other source of income is available.

Since 1996, charcoal production is only allowed for the clearance of farmland. In the past, about 40 km² of savannah woodlands were affected annually by the production of charcoal (Orgut AB, 1995 a).

In 1979, under the terms of a Dutch-Tanzanian joint-venture, the first mechanised farm for multiplication of bean-seeds opened in Naberera. It was economic liberalisation in the early 1980s that brought about new investment opportunities with urban investors seeking land and entering 'grower-contracts' with the seed companies. At present, about 7,000 ha are cultivated each year for the production of seed-beans, beans, wheat and maize in Naberera. More of the land, the exact amount of which is not known, is allocated (or was taken illegally) and awaits further development. In 1996, Ilaramatak Lokoneroi, a Maasai run Non-Governmental Organization (NGO), took twelve cases of illegal land acquisition to the court. However, none of the cases pending affects local Maasai leaseholders.

These farms offer an opportunity for local smallholders to rent tractors and other machinery. Rich Maasai households can afford to cultivate farmland of about 30 to 50 acres mechanically and to hire labour for weeding. In poorer households, apart from soil preparation all work is done by members of the family.

The development of farmland reduces grazing resources. About 7,000 ha of grazing land, though partly infested by tsetse flies, were lost directly to the farms. A more severe consequence is the inaccessibility of important seasonal and farm-based waterponds. Using map analysis, it can be shown that about 30,000 ha of grazing land can presently be used for a period of up to three months only, while this had been six months before the establishment of farms (Schade and Ibrahim, 1999). This happened although farmers claim to respect the needs of local resident Maasai. In reality, farmers turn out to have no knowledge about the Maasai land use system with hardly any social contacts existing between the two groups of land users. Pastoralists see themselves urged to use tsetse infested land for grazing and, thus, are very much inclined not to oppose the destruction of tsetse habitat by charcoal production.

Research Design Used

Many integrated rural development programs having a natural resource management component are built upon district structures for local support while the focus is on the village being the lowest administrative unit of program implementation. The hypothesis put forward here is that the focus upon village level is misleading since resource management decisions in pastoralist societies are (still) not made on village level. Furthermore, it could be stated that the current change in land use is accelerated by the abuse of power by local Maasai members of the village council who could no longer be held responsible for their deeds since they have already left the traditional institutions.

Concerning regional data availability, many baseline data for Naberera area and the pastoralist economy were found to be available in unpublished reports of development agencies and NGOs. Therefore, no further structured household survey was done. The main research tool applied here were semi-structured interviews which were done with all members of the village council, village officers, charcoal producers and selected farm managers.

In addition, group interviews were done with methods adopted from 'Participatory Learning Methods' (Institute for Environment and Development, 1994). A common feature of the methods used are results to be visualised by simple means like stones, differently coloured earth, sticks and seeds. The visualised results were then discussed by the group in an open manner with the researcher holding the position of a moderator.[3] Group discussions were recorded and tapes, photos and other documents analysed by the researcher and interpreters having a strong cultural background. Thus, the findings presented here will at best be of a semi-quantitative nature. While this poses limitation upon statistical analysis, the assumed disadvantage is seen to be outweighed by far by additional information derived from group discussions.

As an additional research tool, digital map analysis was used in order to measure the area of land lost directly or indirectly to farming (Eastman, 1997).

Making Decisions on Resource Management

Resources and Decisions in Pastoralist Societies

The anthropological literature on Maasai agreedly states (Ndagala, 1992; Potkanski, 1998) that all households negotiate access to resources with households already residing at a prospective settlement site. Also, it should be part of official practice in district administration that movements of herds across district borders are allowed and local village councils to be consulted for a site to settle and establish a small farm. Village councils could further set up 'village by-laws' on land use that might restrict pastoral or agricultural activities to certain areas.

In Table 6.1 results are presented which show that traditional decision making on resource use and maintenance did not change significantly in Naberera. Wells are owned by the clan or by an individual if he detected the well and invested labour in its development (Potkanski, 1998).

Although wells and dams provide the same resource, the bodies that regulate their use and maintenance differ. A clan meeting and the neighborhood decide upon wells while in the case of dams a meeting of elders from all clans living in Naberera regulate use. Surprisingly, the village council is not concerned with the maintenance of the small dams. Discussions revealed that the village council would be approached if a larger amount of money is needed, for example for repairs of a borehole.

Table 6.1   Resources and institutions in Naberera's pastoral society

	Neigh-borhood	Clan at locality	Section Oloho	Locality Orkong	Olaibon - Almalal	Village Council	Village Assembly	CCM*
Well maintenance	6	20						
Dam maintenance	10			20				
Cutting of trees for own demand								
Calf pasture (*alalili*)	20			5				
Small fields for Maasai	20					5		
Land for large farms						20	5	
Burning of dry grass (*alalili* only)	10							
Charcoal production	30					20		
Fighting tsetse	5			15	15			

* Chama cha mapinduzi (Party of National Unity)

The use of pasture is entirely decided upon at a neighborhood level. The elders living close to one another discuss the establishment of *alalili* which is an area reserved for grazing calves and sick cattle near the settlements during the time when the herds are on pasture further away. Small fields are established without consultation of the village council and will not be opposed if they are outside *alalili*. In the best of all cases the village council is informed about the decisions. The burning of dry grass is only discussed if a drought is expected. Otherwise, grass burning and the use of trees for own demand are not discussed at all and only regulated by customs and beliefs.

Charcoal production turns out to be highly problematic in Naberera area, and the district has already announced to prohibit further production.

Some Maasai neighbourhoods were in favour of charcoal production as it eliminates the habitat of tsetse flies, opens new grazing grounds and, thus, partly releases the pressures upon shrinking pastoral resources. Other neighbourhoods not affected by tsetse, however, were supportive of the prohibition since they were afraid that 'trees cut will take away the rain'. The village council discussed the issue but did not come to a conclusion before the district raised the ban.

Officially, the village assembly has to discuss matters relating to large leaseholds. In all group discussions the point was raised that in Naberera the village council (or the village chairman and his secretary) make the decision and inform the village assembly afterwards. Neither the locality meeting nor the clan meeting are approached. Frequently, the sub-village chairman (i.e., the former 'ten-cell leader') of areas affected was found to oppose the decision openly (Igoe and Brockington, 1999).

Acceptance of Institutions by User Groups

According to Tanzanian law, the village council, the village chairman and district authorities hold the power to regulate resource use. But these local governments lack the ability to control regulations. Furthermore, the question arises whether they are accepted by the users.

In order to investigate the relative powers of different institutions, a mention of each institution was given to respondents. They were asked whether an institution 'influences or might influence his decision where to put the herds', 'where to cut trees' or 'take a piece of land out of production'. Institutions not given a mention by the respondents were eliminated from the list, remaining institutions ranked in the order from 1 to 5, and point scores added (e.g., 3 points for rank 1 and 2.5 points for rank 2, down to 1 point for inclusion in the list of influential institutions). Allowing for a breakdown by groups of users (i.e., farmers, charcoal producers and pastoralists), in Table 6.2 results are specified in terms of the percentage achieved as compared to the maximum number of points possible.

The findings of Table 6.2 indicate that farmers are mainly influenced by the experience of other farmers and by Maasai living near the farm. Farmers seem to accept the village council as well as village by-laws. However, it was found that the 'co-operation' with the council only relates to complains that herders enter farms illegally. The village council is the body that is approached for donations and voluntary contributions to village development.[4] As the district lacked in 1996 the means to control regulations and any device to oversee whether farms occupy land according to their title deeds, the District Authorities (Commissioner, Council and Land Management Programme) are not regarded as important by the farmers.

Table 6.2   Acceptance of institutions by user groups

Authority/Organisation	Farmers (n=4)	Charcoal producers (n=20)	Herders (n=24)
Regional officers		31.0	11.1
District Commissioner	22.2	53.4	48.1
District Council		24.1	18.5
Village council, village by-laws	50.0	44.8	27.8
Village Chairman		30.1	18.5
LAMP* (donor funded)	27.8	31.9	27.8
Church based NGOs (e.g., ADDO*)		57.8	29.6
Churches in Naberera		39.7	22.2
Traditional spiritual leaders			61.1
Maasai NGOs (*Ilaramatak Lokoneroi*)		14.7	25.9
(other) Charcoal producers		8.6	
(other) Farmers	77.7	15.5	3.7
Maasai-Pastoralists living nearby	61.1	24.1	66.7
Family, relatives (charcoal producers)		58.7	
Sons, *Ilmurran* (Maasai families)			75.9
Wives (Maasai families)			24.1
Traders	22.2	12.1	

* ADDO = Arusha Diocese Development Organisation
   LAMP = Land Management Project of Simanjiro District

Surprisingly, charcoal producers mention that their families have the strongest power on them to stop charcoal production. This implies that protection of woodlands can only be achieved by programs addressing the families in the respective home areas of charcoal producers. The influence of a church based NGO was highly ranked because it was seen as the only organisation addressing the problem of water supply in Okutu being a charcoal makers' settlement that intends to become a registered village. Also, local churches were given high ranks in terms of nearly 40% of the votes. During the discussions it became clear that there are hardly any contacts between the migrant charcoal makers, the pastoralists and the village administration. In locational terms, churches and the bus stop turn out to be the only places where Maasai and charcoal producers meet one another.

Most of the herders interviewed were at the same time members of the village council, but do not attribute strong power to the village council supposed to regulate the use of pastoral resources. Instead, they tend to accept the advice of their sons (*Ilmurran*), of Maasai in the neighborhood and of traditional spiritual leaders. The district commissioner has statedly some power, while other institutions turn out to have only minor influence.

Obviously, there exists a difference between the self-assessment of user groups (about influential institutions) and the perception of village government members – see Figure 6.3. Village council members regard district authorities to be more powerful than they are in the perception of charcoal makers and farmers. Council members assess their limited role properly, but seemedly are not aware of the power given to them by law.[5]

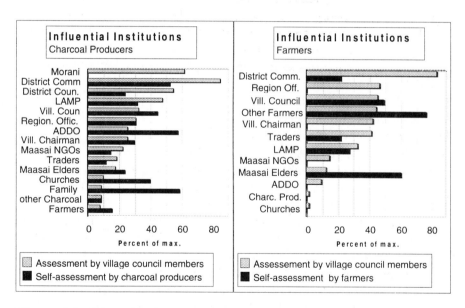

Figure 6.3 Diverging patterns of institutional (self)assessment

The Village Council: A Representative Organ?

The village council of Naberera proves to be not a homogenous body. In order to identify persons of particular influence, all council members met during the time of fieldwork were asked with whom they discuss matters of the village council informally more than twice a month. The results of an informational network analysis are given in Figure 6.4 with dark boxes indicating close contacts within sub-groups of the council members sampled.

The matrix is not symmetricised so that the information 'village council member no. 1 (fifth column) contacts no. 16 (7th row)' not necessarily implies that 'council member no. 16 is in contact with no. 1'. In the second column of the matrix the residential place of council members in the Naberera's settlement clusters ('subvillages') is given, being specifically the central village (Cen), Esoit (Eso), Kidomungen (Kid), Kosiki (Kos), Olosoito (Olo) and Soito (Soi). Only for male Maasai specifications of age-set memberships and positions in the administration are provided in the third column (age-set), i.e., village chairman (VC), village secretary (VS), *Seuri* (S), *Makaa* (M) and *Landis* (L) (*Landis* formed the *Ilmurran* age-class in 1996/97). Other specifications in the third column are (Wom) for female members of the village council and (Off) for appointed non-Maasai Officials working in Naberera. State officers are not full members of the village council but of the respective committees for health, education etc. Maasai clan affiliations – but also immigrants ('Swa' for Swahili people) – are specified in the fourth column, i.e., Mollel (Mo), Laizer (L), Markassin (Mr), Mamasita (A) and Laitok (Lj).

The data prove that the informational flows are mainly determined by the village council members' place of residence. The network is closest among those village council members living in the village centre and among those living in the subvillages of Kidomungen or Kosiki. It is not as tight between Maasai members living in the centre and non-Maasai living there. It is very loose between people living in the centre and those living outside and those living farther away at different places.

The close informational ties between the village council members in the centre bring about as a conclusion that just a few people hold important positions. When, for instance, the Arusha Diocese Development Organisation (ADDO), a church based development NGO, came to the village promoting environmental village by-laws, a rather small (core) group of centre based council members drafted the laws and handed them directly to the district council for approval without consultation of any other council members or the village assembly (as laid down by the Local Governments Act of 1982).

The same core group again took part in interviews when the donor-funded Land Management Programme of Simanjiro District (LAMP) collected data for a baseline study (Orgut AB, 1995 b). Thus, it is not

Subv.	Ageset	Clan	18	13	17	8	6	2	12	11	10	7	22	30	29	27	26	24	16	20	15	5	4	3	1
1 Cen	S/VC	Mo	x			x		x	x			x	x							x	x	x	x	x	x
3 Cen	S	Mr	x	x		x	x		x		x	x	x	x	x	x	x	x		x	x	x	x	x	x
4 Cen	M	A					x	x				x	x	x	x	x	x	x		x	x	x	x	x	x
5 Cen	M	Mo			x	x	x	x				x		x	x	x	x	x		x	x	x	x	x	x
15 Cen	Wom	Mo	x	x		x	x	x				x		x	x	x	x	x		x	x	x	x	x	x
20 Cen	S	Mo				x		x				x	x	x	x	x	x	x		x	x	x	x	x	x
16 Cen	Wom	Mo	x		x		x					x	x						x	x	x		x		x
24 Cen	-	Swa				x	x				x		x	x		x	x	x		x	x			x	x
26 Cen	Off	Swa	x		x	x	x	x				x	x	x	x	x	x	x		x	x	x		x	x
27 Cen	Off	Swa			x		x	x				x		x	x	x	x	x		x	x	x		x	x
29 Cen	Off	Swa	x	x	x	x	x	x					x	x	x	x				x	x	x		x	x
30 Cen	Off	Swa				x	x					x	x	x	x	x	x	x		x	x	x	x	x	x
22 Eso	Wom	A	x	x								x	x						x		x				
7 Kid	S	A	x		x			x	x	x	x	x	x	x	x	x	x	x		x	x	x	x	x	x
10 Kid	S	L			x	x	x	x	x	x	x	x	x	x	x									x	
11 Kid	S	L				x		x	x	x	x	x	x												
12 Kid	L	L	x	x		x	x	x	x	x	x	x	x						x						x
2 Kos	L/VS	Mo			x	x	x	x	x				x									x	x	x	x
6 Kos	M	Mo	x	x		x	x	x	x			x	x									x	x	x	x
8 Kos	L	Mo		x	x	x	x	x	x			x	x					x				x	x	x	x
17 Nos	M	Lj			x								x												x
13 Olo	Wom	Mr	x	x		x	x	x					x												x
18 Soi	M	A	x		x	x		x				x	x									x		x	x

First column: number of respondent

Second column: Subv. = Subvillage / place of residence within Naberera Cen (Centre), Eso (Esoit), Kid (Kidomungen), Kos (Kosiki), Olo (Olosoito), Soi (Soito)

Third column: Age-set memberships and positions held in village administration: S (Seuri), M (Makaa), L (Landis), Wom (Woman), Off (Village Officer), VC (Village Chairman), VS (Village Secretary)

Forth column: Clan affiliation Swa (Non-Maasai), Mo (Mollel), L (Laizer), Mr (Markassin), A (Mamasita), Lj (Lajtok)

The shaded areas indicate close contacts within sub-groups of the sample.

Figure 6.4   Informational network analysis of Naberera village council

surprising that these members also hold 'rights of occupancy' for plots of 500 or 1000 acres (200 or 400 ha) looking for investors or handing land to outsiders for cultivation.

Despite of this, the village council might be a valuable institution if strong personal ties to the traditional institutions that *de facto* regulate resource use existed.

Village Council Members and Traditional Meetings

To acquire data on the co-operation of influential village council members and traditional meetings, an experiment was done that resembles the 'wealth ranking' exercise from the PLA-toolbox. The names of all male *Ilmurran* and Elders from two sub-villages, those of all male Maasai village council members, those of former age-set leaders and the names of other highly influential Maasai from the village were written on cards. Separately, respondents were asked to group the names of persons that were seen to have 'the same influence and power in the traditional meeting'. One by one was read the names on the cards were read and the respondent could himself decide how to group the names. In addition, the respondent could also exclude names if he did not know the person, or if the person did not participate in the meetings. When all persons were grouped the respondent was asked to explain the characteristics of the groups.[6] The results of 'power ranking' turned out to be in harmony with previous findings and can be summarised as follows:

- most village council members participate in the traditional meetings, but the core group does so quite seldom;
- the core group are considered 'people who like meetings but don't accept decisions', 'they change a good idea to a bad idea for their own profit', 'they divide people', and 'they have their own meetings, where they don't invite anyone'; they were looked upon as being the weakest group in the meeting;
- the village council members living outside the centre (and following a traditional life style) together with men included in the exercise because of their wealth were ranked higher than average;
- although the age-set leaders of the *Ilmurran* retire after they have become elders, they remain highly influential; people identified to be most powerful in the meetings were either age-set leaders at localilty (*enkutot*) level or leader's assistants in Naberera;
- the current age-set leader of the *Illandis* is the person ranked highest from *Ilmurran* age-class;

- a local assistant to the age-set leader of the *Ilmakaa* (i.e., the age-set preceding the *Illandis*) was also village chairman in the past; he was accused of corruption and of giving away land illegally to farmers; he was first taken the stick (being the symbol of the age-set leader), but continued to act as village chairman; later he was voted out of office by a village assembly meeting; his present influence in the traditional meeting is very weak;
- although some members of the *Ilmurran* age-set have an influential voice in the traditional meeting, the majority of them are considered to be 'too young' or 'inexperienced'; and
- very old men rarely participate; but if they do, they are considered to be very strong advisors.

When asked why the most influential people in the traditional meeting were not members of the village council, two points were raised. First, 'they are very busy because they are consulted by many people'. And second, 'they don't regard the village council as an important body to make decisions'. It was further suggested that the self-nomination as representative does not go hand in hand with traditional Maasai electoral rules. Traditionally, a special meeting of Maasai elders will discuss and take an unanimous decision about an *Ilmurran* age-set leader without the potential candidate being asked in advance. Neither the modern nomination process nor the majority rule is seen to accord to traditional principles.

Discussion of Results

Power and Changes of Land Use and Institutions

Schumpeter (1964) stated that the innovative entrepreneur is the engine of the development process, and North (1990) describes how institutional reform takes place as a result of a political process after the relative prices of resources in the economy have changed. Political power relations are seen by North (1990) to constitute an important determinant of shaping future institutional arrangements. In Naberera, both of these aspects can be identified – and both will provide an incomplete picture.

Ujamaa-policy and village land registration introduced powerful local councils and party representatives (Shivji, 1995), but it did not set up effective controls of these powers (Bruce, Freudenberger and Ngaido, 1995).

The state was successful in establishing its control over land in a top-down approach by creating national parks, game reserves, state owned ranches and military land. This took away – and still continues to do so (Brockington, 1998) – important resources from local communities that have to struggle harder than in the past to make a living in the drylands.

Technological progress such as improved maize varieties and machinery became available locally through government supported initiatives (i.e., the opening of a farm for seed multiplication) that originally did not put people in the centre of developmental activities. New investment opportunities attracted the urban *élite* looking for land in order to copy the successful farming initiatives.

At the local level, those who had knowledge about the on-going reforms (e.g., party members and relatives of a local member of parliament), tried to use their power for their own benefit. In the sense of Ribot (1998), the most important determinant of access to benefits was information and a powerful position within the institution that *de jure* had to be approached for a title deed, although the village council *de facto* did not regulate resource use. Various strategies of profiting from their position of powers can be observed in Naberera village. First, bribes are accepted so as to forge council minutes (which was done at the neighbouring village of Engasumet) and give land to outsiders. Second, land was allocated to oneself with farmers recruited and requested to agree to sharecropping contracts. Third, some ten hectares or so are cultivated with the help of local farmers who need to stay in good terms with the village council.

Economic liberalisation brought about a change in the relative prices of resources that North (1990) considers to be the starting point of institutional reforms. However, it could be challenged whether institutional reforms happen.

With regard to the traditional system of land use, only slight changes could be observed that relate to new opportunities arising from small-scale farming. Maasai settlements consist of a ring of houses and a place for the cattle inside this circle. Over the years manure accumulates in the inner circle. When the settlement is abandoned (mainly because of parasites accumulating), the whole settlement is burnt. Today, the former owner of the place has to be asked whether he wants to cultivate the place. It used to be a place free for any new settlement in former times. No institutional reforms, however, have occurred in the administration of water resources and trees.

With regard to the local administrative system, reforms turn out to have occurred only to a limited extent. As the core group of village council members rarely participates in the traditional meeting, this meeting can no longer exert pressure upon them. Institutional innovation is brought to the village administration from outside by the district administration and the NGOs working in the area. Both of them get in contact with the core group faster than with ordinary members. The draft of the environmental by-laws put the heaviest burden on charcoal producers and the 'Swahili households residing in the village centre'. They were expected to pay for licenses for the collection of fuelwood and the cutting of trees in the vicinity of the village. This would not have much impact on the pastoralist community since nobody intended to control the collection of wood farther away.

The by-laws were intended for a separation of grazing and farming areas. Naturally, all the title deeds of the core group lay in the prospected farming areas.

In summary, effective institutional reforms relating to the use of resources were relatively minor. However, a separation within the Maasai society took place. On one side, there are Maasai who still practice animal husbandry as the main source of income, cultivate small fields (sometimes with the help of employed labour) and rely on acceptance, membership and negotiation in the traditional institutions when seeking access to resources. On the other side, there are those who know where the legal power is to regulate resource use. This situation is similar to districts of southern Kenya where power and access to information determines individual success during the privatisation of group ranch land (Galaty, 1994; Bekure, de Leeuw, Grandin and Neate, 1992).

Two main effects of land use changes in Naberera need to be briefly discussed, i.e., the impact upon traditional insurance systems against drought and upon natural resources and game habitat.

Impact on Traditional Insurance Systems

Potkanski (1998) describes the institutionalised system of restocking poor pastoralists after drought or animal epidemics. Individual households are restocked by donations from friends and relatives, and in cases of severe loss by a joint effort of the clan.

In a dynamic modelling approach, Mace (1993) showed that the most successful livelihood strategy of dryland households having a choice between herding and farming activities will depend on their initial wealth. Poor households have to rely on farming activities, while mobile animal husbandry turns out to be the most successful strategy for richer households. Following and modifying this, it could be stated that an optimal strategy for rich households actually is the investment of a part of the herd in farming activities and the investment of farming surplus in animal husbandry. Naberera village council's core group is seen to currently practise this type of land use strategy.

Large-scale farming, in general, is not an option for poorer households. The investments are too high and the failure of rains too likely. However, in Naberera most of the households practice farming nowadays. Due to failure of rains and investment in agriculture, some households live on the edge of destitution (Ibrahim and Ibrahim, 1995). They will become entitled to restocking by the clan, but as some very rich households (i.e., members of the village council's core group) left the traditional institutional system, livestock resources for an efficient restocking will be limited.

A large portion of households will have to rely on other sources of income such as selling traditional medicine and gemstones or using up remittances from *Ilmurran* working in towns (Brockington, 1998; Schade and Ibrahim, 1999) in future.

Impact on Natural Resources

Results from the Serengeti (Dublin, Sinclair and McGlade, 1990) suggest that East African savannahs are ecosystems with multiple equilibria. The combined effect of rainfall, grazing and fire in the last decades determines whether the ecosystem is today dominated by trees or grasses. Under these conditions degradation cannot be proved from vegetational analysis alone (Smith, 1995). Instead, the most important indicator for degradation is the condition of the soil.

In Naberera, soil and vegetation of grazing land are in a 'surprisingly good condition' (Kahurananga, 1996). In contrast to this, wind and water induced soil erosion can be observed on farming land (Lehmann, 1998), in particular on land that was illegally acquired (Schade and Ibrahim, 1999). As a matter of fact, the willingness to respond to and cope with soil degradation depends on the legal status of the land title as well as upon the individual farmer's knowledge. While, for example, the oldest farm in Naberera has shifted cultivation practices to a no-till system that allows to rebuild soil organic matter, at the same time, some of the 'new farmers' from urban areas don't even practice management of soil nutrients. Yet, there are no studies of succession on abandoned fields being depleted of resources and soil organic matter. In the worst of all cases, unpalatable species of grasses and bushes will dominate these areas in future.

During the last decade, some hundreds of square kilometres of wooded savannah have been depleted of trees due to the production of charcoal on Naberera village land. These woodlands were an important wildlife habitat used for tourist hunting activities and supporting Tarangire National Park as a destination area for migrating game (Borner, 1982; Tarangire Conservation Project, 1996).

The game industry has shown an interest in protecting the resources, but local Maasai communities hardly benefit from tourism related activities. The villages' shares from 'animal head fees' derived from hunting were not distributed by district authorities. Quarrels about donations from tour operators to village councils are frequent (World Bank, 1997). The community conservation programme of the National Park Authority (Tarangire National Park, 1994) has been put into operation but experienced considerable problems due to the bad perception of the National Park as a consequence of older proposals aimed at the establishment of a 'conservation area' that restricts farming activities around the park (Borner, 1982).

In the near future, interest in wildlife resources will be crucial for the development of the Maasai steppe. Farming activities and protection of game are incompatible. Animal husbandry activities and interest in game protection are compatible as long as the strong Maasai behavioural taboos not to hunt game still hold. This importance of taboos for the protection of species had recently been emphasized by Osemeobo (1996) and Colding and Folke (1997). However, with regard to the area studied, behavioural taboos are assumed to change as soon as the interests of Maasai household in farming activities rise and the protection of fields becomes a necessity for survival.

Conclusions

Land use changes in Naberera village area have proved to be a source of conflict as it could be seen from a number of incidents between farmers and pastoralists during the 1996/97 drought. Pastoralists lost more land due to the inaccessibility of important watersources than due to the conversion of grazing areas into farmland. This is a result of lacking social contacts between the farmers and the pastoralists concerned. Whereas farmers follow the 'legal' way, i.e., regard the village council as the representative body for the Maasai community, Maasai pastoralists regulate the use of grazing and watering resources either in neighborhood meetings or by operating on a much larger spatial scale. While in case of problems occurring, farmers address the core group of village council members living in the village centre, core group members cannot serve as contact people for the Maasai community.

Both groups of land users were found to be ignorant and full of prejudices about each other. This is at the root of most problems existing. Development agencies and NGOs should address this problem by moderating contacts between farmers and neighbourhoods to discuss and solve problems. In other African regions synergetic effects between herders and farmers exist, and through mediating contacts, development programs might yield positive effects.

Only the participation of a large part of society can accelerate necessary institutional reforms. But participation requires the ability to participate and as in a well organised hiking group guides will have to find a compromise between the speed of the weakest and the speed necessary to reach the mark. In the past and even today, farmers and state administrators moved too quick for the majority of the local community.

By addressing neighbourhoods and the traditional meetings, farmers and administrators might find a way of not being dependent upon village councils and softening the negative effects of institutional reforms upon a marginalized group such as the pastoralists – without eliminating assumed positive effects emerging from the introduction of improved land use technology.

According to Serageldin (1993), it could be stated that the cultural dimension is the forgotten third pillar of sustainable development. Institutions, be it in the form of norms, taboos or written laws, are the base of people's interaction. Every new technology will lead to institutional reform. Whereas the individual's access to new technology will depend on economic and political power, the resulting external effects from technology on society depend on people's participation in institutional reforms. In Naberera, the benefits of farming for a few were so big that they no longer have an incentive to share the institutions of the Maasai community. In the long run this might lead to the disintegration of the community.

Notes

[1]   Personal communication by Mr. Bruinsma, farm manager of Royal Sluis Farm, Naberera in April 1997.

2   'Section' has no particular Maasai language equivalent. The corresponding traditional term is *olosho* meaning 'plateau'; see Ole Ngulay, undated, p. 34.

3   An example is provided in Table 6.1. The figures given there indicate the average number of seeds placed in the respective field of the matrix during the group discussions on decision-making institutions. Large numbers imply that 'decision is made here', while small numbers relate to bodies 'discussing the way how work is carried out'.

4   Farmers contribute voluntarily to village development, for example, by financing school buildings or by installing a solar driven waterpump. Although the value of donations is considerable for the village, it is minor when compared to profits from farm operations, i.e., about US$ 0.13 per ha of farmland on a yearly basis (Schade, 1997). Quarrels about donations (and their misuse) are reported from Naberera as well as from neighbouring villages where the National Park Authority and/or tourist agencies contribute to village development.

5   Unfortunately, charcoal producers were not asked about the influence of the *Ilmurran* (sons of the herders). When interviewed, Maasai village council members identified the *Ilmurran* as most powerful because 'they beat the charcoal makers, then they leave everything behind and go'.

6   It has to be mentioned that results were obtained from only three respondents since the interpreter was approached by a highly respected elder who heard about the exercise and felt at unease that his name was treated in that way.

References

Bekure, S., de Leeuw, P.N., Grandin, B.E. and Neate, P.J. (1992), *Maasai Herding: An Analysis of the Livestock Production System of Maasai Pastoralists in eastern Kajiado District, Kenya,* International Livestock Centre for Africa, Addis Abeba.

Borner, M. (1982), *Recommendations for a Multiple Land Use Authority adjacent to Tarangire National Park, Arusha Region Tanzania,* Frankfurter Zoologische Gesellschaft, Frankfurt/M.

Brockington, D. (1998), *Land Loss and Livelihoods: The Effects of Eviction on Pastoralists moved from the Mkomazi Game Reserve,* University College London (PhD Thesis).

Bruce, J.W., Freudenberger, M.S. and Ngaido, T. (1995), 'Old Wine in New Bottles: Creating New Institutions for Local Land Management', in Deutsche Gesellschaft für Technische Zusammenarbeit (ed) (1997), *Bodenrecht und Bodenordnung: Ein Orientierungsrahmen,* Universum, Wiesbaden (Appendix of Individual Workshop Reports on cd-rom).

Colding, J. and Folke, C. (1997), 'The Relation between Threatened Species, their Protection and Taboos', *Conservation Ecology,* vol. 1 (1), online magazine <http://www.consecol.org/Journal/vol1/iss1/art6>.

Dublin, H.T., Sinclair, A.R.E. and McGlade, J. (1990), 'Elephants and Fire as Causes of Multiple Stable States in the Serengeti-Mara Woodlands', *Journal of Animal Ecology,* vol. 59, pp. 1147-64.

Eastman, J.R. (1997), *Idrisi for Windows, Version 2.0,* Clark University, Worcester.

Galaty, J.G. (1994), 'Rangeland Tenure and Pastoralism.' in E. Fratkin, K. Galvin and E.A. Roth (eds), *African Pastoralist Systems: An Integrated Approach*, Lynne Rienner, Boulder, pp. 185-204.

Holling, C.S. (1973), 'Resilience and Stability of Ecological Systems', *Annual Review of Ecology and Systematics*, vol. 4, pp. 1-23.

Ibrahim, B. and Ibrahim, F.N. (1995), 'Pastoralists in Transition: A Case Study from Lengijape, Maasai Steppe', *GeoJournal*, vol. 36 (1), pp. 27-48.

Igoe, J. and Brockington, D. (1999), *Community Conservation, Development and Control of Natural Resources: A Case Study from north-east Tanzania*, Pastoral Land Tenure Series No. 11, Institut for Environment and Development, London.

Institute for Environment and Development (1994), *Special Issue on Livestock*, RRA Notes No. 20, IIED, Sustainable Agriculture Programme, London.

Inyuat-e-Maa (1991), *First Maa Conference on Culture and Development*, Inyuat-e-Maa, Terrat, Arusha.

Kahurananga, J. (1996), *Rangeland Survey for the Simanjiro Land Management Project*, Land Management Project of Simanjiro District, Engasumet.

Land Commission (1994), *Report of the Presidential Commission of Inquiry into Land Matters*, United Republic of Tanzania, Ministry of Lands, Housing and Urban Development, Dar es Salaam.

Lehmann, J. (1998), *Report to Tropical Ecology Support Program*, Department of Soil Science and Soil Geography, University of Bayreuth.

Lindberg, C. (1996), *Society and Environment Eroded: A Study of Household Poverty and Natural Resource Use in Two Tanzanian Villages*, Geografiska Regionstudier No. 29, Uppsala University.

Loiske, V.-M. (1995), *The Village that Vanished: The Roots of Erosion in a Tanzanian Village*, Department of Human Geography, Stockholm University.

Mace, R. (1993), 'Transition between Cultivation and Pastoralism in Sub-Saharan Africa', *Current Anthropology*, vol. 34 (4), pp. 363-82.

National Environment Management Council (1993), *Baseline Mapping for Monitoring of Desertification in Naberera*, National Environment Management Council, Dar es Salaam.

Ndagala, D. (1992), *Territory, Pastoralists and Livestock: Resource Control among the Kisongo Maasai*, Uppsala Studies in Cultural Anthropology No. 18, Stockholm.

North, D. (1990), *Institutions, Institutional Change and Economic Performance*, Cambridge University Press.

Ole Ngulay, S.O. (undated), *Kisongo Maasai Customary Land Tenure Arrangements: Social and Ecological Considerations for Sustainable Pastoral Livelihoods*, Institut for Environment and Development Paper, London.

Orgut AB (1995 a), *Environmental Impact Assessment Study of Simanjiro District, Arusha, Tanzania*. Orgut Ab, Stockholm, Dar es Salaam.

Orgut AB (1995 b), *Ukusanyaji wa Taarifa za Awali za Hifadhi Ardhi na Mazingira, Januari, 1995, Wilaya ya Simanjiro, Kijiji ya Naberera*, Land Management Project of Simanjiro District, Engasumet.

Osemeobo, G.J. (1996), 'The Role of Folklore in Environmental Conservation: Evidence from Edo State, Nigeria', *The International Journal of Sustainable Development and World Ecology*, vol. 1 (1), pp. 48-55.

Potkanski, T. (1998), *Pastoral Economy, Property Rights and Mutual Assistance Mechanisms among the Ngorongoro and Salei Maasai of Tanzania*, Pastoral Land Tenure Series Monograph No. 2, Institute for Environment and Development, London.

Ribot, J.C. (1998), 'Theorizing Access: Forest Profits along Senegal's Charcoal Commodity Chain', *Development and Change*, vol. 29, pp. 307-41.

Rigby, P. (1988), 'Class Formation among east-african Pastoralists: Maasai of Tan-zania and Kenya', *Dialectical Anthropology*, vol. 13, pp. 63-81.

Schade, S. (1997), 'The Meaning of "efficient resource use": The case of Northern Tanzanian Drylands', in F.N. Ibrahim, D. Ndagala and H. Ruppert (eds), *Coping with Resource Scarcity*, Bayreuther Geowissenschaftliche Arbeiten No. 16, University of Bayreuth, pp. 117-57.

Schade, S. and Ibrahim, F.N. (1999), *Nachhaltigkeit konkurriender Landnutzungssysteme in der Baumsavanne Nord-Tansanias: Ursachen, Begleitumstände und Folgen des Landnutzungswandels im Simanjiro Distrikt*, Tropical Ecology Support Program Report, Deutsche Gesellschaft für Technische Zusammenarbeit, Eschborn.

Schumpeter, J. (1964), *Theorie der wirtschaftlichen Entwicklung*, Dunker & Humblot, Berlin.

Serageldin, I. (1993), 'Making Development Sustainable', in I. Serageldin and A. Steer (eds), *Making Development Sustainable: From Concepts to Action*, Environmentally Sustainable Development Occasional Papers Series No. 2, World Bank, Washington DC, pp.1-5.

Shivji, I.G. (1995), 'Land Tenure Problems and Reforms in Tanzania', in Deutsche Gesellschaft für Technische Zusammenarbeit (ed) (1997), *Bodenrecht und Bodenordnung: Ein Orientierungsrahmen*, Universum, Wiesbaden, Appendix of Individual Workshop Reports on cd-rom.

Smith, E.L. (1995), 'New Concepts for Assessment of Range Condition', *Journal of Range Management*, vol. 48, pp. 271-82.

Tarangire Conservation Project (1996), *Analysis of the Migratory Movements of Large Mammals and their Interaction with Human Activities in the Tarangire Area in Tanzania as a Contribution to a Conservation and Sustainable Development Strategy*, University of Milano, Varese.

Tarangire National Park (1994), *Community Conservation Service Action Plan*, Proceedings of a workshop held in September 1994 at Tarangire National Park.

United Republic of Tanzania (1991), *Census Data 1988*, Bureau of Statistics, President's Office, Planning Commission, Dar es Salaam.

Waller, R. (1985), 'Economic and Social Relations in the Central Rift Valley: The Maa-speakers and their Neighbours in the 19th century' in B.A. Ogot (ed), *Kenya in the 19th Century*, Bookwise, Nairobi, pp. 83-151.

World Bank (1997), *Benefit Sharing in Protected Area Management: The Case of Tarangire National Park, Tanzania*, Findings: African Region No. 88, World Bank, Washington DC, <hhtp://www.worldbank.org/aftdr/findings/english/find 88.htm>.

7   Drought Hazards and Threatened Livelihoods
Environmental perceptions in Botswana

FRED KRÜGER

Introduction

Drought is a major risk factor for sustaining livelihoods in Botswana. Statistically, one in three years is a drought year. This figure, of course, only roughly paraphrases the true impact of drought events on society and livelihood systems. Particularly devastating are several droughts that follow in succession (for example, 1981/82 to 1986/87) or droughts that include extremely high temperatures and cover larger parts of the southern African sub-continent. Botswana still has not entirely recovered from the latest, and, in living memory, worst drought cycle which began in 1991/92 and affected all of southern Africa – see Figures 7.1 and 7.2.

However, and contrary to many Sahelian or even other southern African states, there were no losses of human lives directly related to drought events for the last decades. This is mainly due to very effective governmental drought relief measures, which have gained a lot of attention for their success (Holm and Morgan, 1985; Hay, 1988; Teklu, 1994; Krüger, 1997). On the other hand, a number of shortcomings of these programmes have been revealed lately. There seems to emerge a growing gap between what is officially intended by the drought relief schemes, and what is their actual outcome in the target areas or their true impact on the target groups. As will be shown in this paper, this has a lot to do with the way drought is perceived by governmental institutions (here referred to as 'the State'), and how it is seen and experienced by the people or households who are directly affected or even threatened by it.

In general, drought must not merely be seen as a natural incident involving lack of rainfall and subsequent decline in agricultural plant or animal production. A drought hazard may be triggered by a more or less

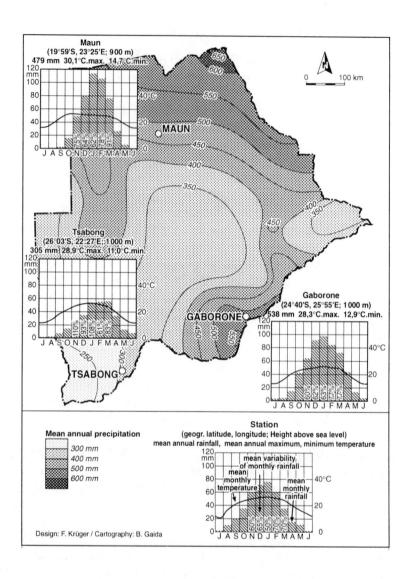

Figure 7.1   Mean annual precipitation and temperatures in Botswana
Source: Department of Meteorological Services

sudden precipitation deficiency, but it also always includes disruptions of interrelations within and between the physical environment and the social and economic spheres. In short, a drought cannot occur in an arid environment (desert). Under arid conditions the associated socio-economic spheres do not rely on or interact with rainfall. Drought hazards strike under semi-arid to semi-humid conditions, with an average long-term precipitation potentially sufficient for agriculture but a high annual rainfall variability pattern making agricultural activities risky and unsafe. Botswana's semi-humid ecosystem, with temporally and regionally unreliable distribution of rainfall, therefore is highly susceptible to drought, and so are the country's social and economic systems.

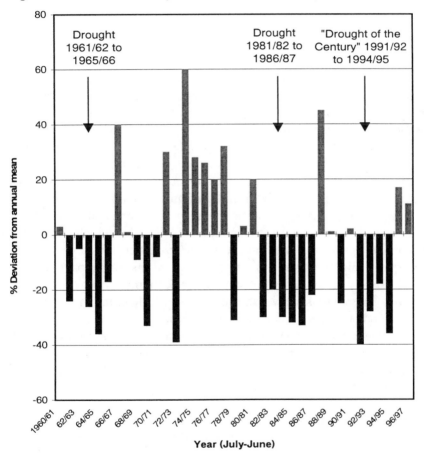

Figure 7.2   Mean annual rainfall in Botswana 1960-1997

Source: Author's calculations based on data from the Department of
                  Meteorological Services, Gabarone

In an attempt to define drought in this broader sense, and to contribute to a better understanding of drought hazards and the creation of effective drought relief measures in Botswana, Sandford (1979) described drought as 'a rainfall-induced shortage between the supply of a good and the requirements for it', a shortage which is 'brought about by inadequate or badly timed rainfall'. This conception helped a lot to design the first drought mitigation schemes in Botswana. The limitations of this concept lie in its narrow focus on the production and supply of goods such as food.

The State's Perception of Drought

As outlined above, drought hazards may not be reduced to rainfall deficiency and subsequent food shortage. A first conceptualization of drought in Botswana was brought about in the early 1970s in an effort by the Botswana government to reduce the negative consequences of drought events and to assist drought-affected people and households. A number of local physical surveys, e.g., by the International Institute for Aerial Survey and Earth Sciences (1981), and various studies on the social aspects of drought, e.g. Campbell (1979), formed a framework for the implementation and later then constant modifications of governmental drought relief measures. This led to a perception of drought which was clearly economically biassed. The point was that no matter how severe a rainfall inadequacy might be from a physical viewpoint, its true impact could only be assessed by how heavily it disturbed the economic system of supply and demand. The obvious answer to drought events, then, lay in the attempt to bolster food production in order to support the supply side of the system.

Famine, under- or malnutrition were basically perceived as a result of drought-induced crop and meat production failures. Consequently, some of the major targets of the 'National Food Strategy' (NFS) as set up in the Republic of Botswana (1985; p. iv) were

- increases in production of basic cereals and seeds, in order to increase rapidly Botswana's degree of self-sufficiency particularly in maize and sorghum,
- increases in irrigated land, and
- reductions in the proportion of national food supplies made up from imports.

The NFS was adopted by the Botswana government in 1985 and lay the political groundworks for all national drought relief programmes. After a couple of years, though, it became quite obvious that the target of national food self-sufficiency could never be achieved. The only surplus food product was meat. Domestic food grain production was too low and even after good yields met only some 30% of the national demand, while dropping to a mere 5% in drought years (Mokobi and Asefa, 1988). By the end of the 1980s, it was food security rather than food self-suffiency which became the new guideline for drought relief measures. Accordingly, the main focus was shifted from the supply-and-demand side of food to a more diversified scheme comprising a large number of different drought relief and recovery components.

The National Development Plan (NDP) of 1991 clearly distinguished between self-sufficiency and security and set the course for all future drought relief programmes. It is important to note that from 1991 onwards there is also a distinction between supply of food and access to food – a new perception of drought risk and food security which has strongly influenced all relief measures since. The Ministry of Finance and Development Planning (1991, p. 259) put down the difference between self-sufficiency and food security:

National food self-sufficiency implies producing within Botswana sufficient livestock, sorghum, millet, maize, and milled products of these grains to satisfy the food needs of everyone in Botswana, regardless of cost. Given the environmental constraints in Botswana, this cost would be very high. Furthermore, national food self-sufficiency is only indicative of the physical supply of food: it does not guarantee universal access to food, nor the end of hunger and malnutrition. / *Food security*, on the other hand, allows for production and income generation which follows the principle of comparative advantage through trade. The policy of national food security to be implemented during NDP 7 has the following three components:

- at the *national level*, Botswana has a comparative advantage in producing livestock (particularly beef) and sorghum, as well as in the production of minerals and an increasing range of manufactures and services; foreign exchange earnings from activities in which Botswana is competitive can be used for importing essential food items that cannot be produced so advantageously in Botswana, such as maize;
- at the *household level*, it is the purchasing power of individual households which determines the quantity and quality of the food consumed; purchasing power depends both on household income and on the prices of basic foodstuffs; the goal of household food security demands that each household has sufficient income generating opportunities and access to food to meet its nutritional requirements; and

- to lessen risks and dependence, it is wise to buy the imports from a number of countries and to maintain a *Strategic Grain Reserve* in Botswana; currently, it is planned to store up to six months' supply of food grains in such a reserve; this will also help to cushion Botswana against severe shortfalls in production resulting from drought, allowing time to arrange supplies from other sources.

There are two basic concepts immanent to this approach. On the national scale, the aftereffects of drought are seen to disarray a more or less balanced economy of comparative advantage. This must be countered by providing incentives for foreign investment, constant diversification of manufacturing goods and services, and a sound, liberal financial policy. On the household level, a drought event is perceived as a major reason for deteriorating income generating opportunities, the decline of value of productive assets, and subsequent increase of famine risk. This perception of drought follows the approach of 'declining entitlements' (Sen, 1981). It implies that famine not necessarily occurs because of a lack in food supply, but because people lose access to food, meaning they lose the ability to produce, purchase or barter for it. Thus, the negative effects of drought can only be mitigated by transferring income to households facing drought-induced destitution, by preserving their productive assets (e.g., land and cattle), and, in general, by sustaining their livelihoods through a large variety of different relief measures.

On both the national and the household level the major objective is not to provide food, but to combat destitution by trying to alleviate the erosion of entitlements. Supporting the supply side of the market economy alone, therefore, is no longer regarded to be sufficient. The efforts of the Botswana government, based on a democratic political and very liberal economic system, now also involve a set of measures to strengthen household entitlements and back up or encourage demand for goods, even under severe drought conditions. The question is, whether these efforts really meet the 'felt needs' of drought-affected people. In case of a drought, do these people feel they are losing entitlements, or are their actual requirements different from the provisions they receive through governmental relief actions?

How Rural Households Perceive Drought

Obviously, the drought perceptions of the state of Botswana as an institutional body are not necessarily the same as those of drought-affected people who form social groups. Rural households (and it is the rural households who are most directly affected by drought, although – as will be shown below – urban dwellers may also be hit) generally view drought as a trait of life. Household activities, especially those related to agricultural production or food storage, have over decades been customized to alleviate the risk of famine. Not only certain human activities, but also the social value system reflect the immanent importance of rainfall or lack of it. For example, a formal *Setswana* greeting used in ceremonies, welcome addresses and almost all political statements is *Pula*, meaning 'Let there be rain!'. Even Botswana's currency was named Pula to depict the importance of rain, in this case for financial wealth.

The unpredictability of drought events requires a constant variation and adaptation of household strategies in order to cope with drought-induced restrictions. While the state apparatus needs a clear schedule when and where to implement and when and where to stop specific and formalized drought relief measures, households need to be (and can be) more flexible. In terms of economic survival, however, over the past two decades there has been a notable change in rural household structures which allows to better buffer the uncertainties of drought without the need for constant short-term adjustment.

Rural household forms seem to be more and more based on the actual requirements for economic survival, and not so much on kinship-based social organisations (Mazonde, 1999). Of course, the structures of rural households are still very much anchored in kinship relations, but there are clear indications that economic requirements have led to new social values and household organisations which are to a greater extent biassed by economical motives than by 'traditional' kinship norms. It is true that, historically, the risk of drought has always been perceived not only through kinship bindings, but also from a very rational point of view with the overall objective to spread risks (Hitchcock, 1979). Today, though, the sound macro-economic performance of Botswana has opened up a wider range of opportunities to react to the threat of food insufficiency. The transformation of rural household or family structures reflects the weakening influence of kinship relations on decisions connected to drought-induced food shortages.

Households used to view drought as a problem concerning the family, the ward or the tribe. Drought was looked upon as a group problem which was to be solved by consensus within the group. Today, given the quickly modernizing Botswana society with its wider range of coping opportunities, but also a larger importance of individual values and rights, drought is often perceived as an individual problem, with strategies taken up by individuals as outcomes of individually satisfying mediation (Mazonde, 1999). Necessary adaptation strategies and desirable options are now negotiated – if at all – between members of a family or household rather than between members of, e.g., the village or tribal community. Often these negotiations represent a conflict between the pursuit of a family member's personal aims and wishes on the one hand, and the well-being of the household as a social entity on the other.

In rural areas, there are many cases today where families can no longer fall back on household labour which would potentially be available for assistance in food production, e.g., helping out with weeding or field ploughing. Instead, younger family members often opt for waged employment rather than assist their parents in subsistence crop production. Drought is still perceived as a decline in purchasing power and wealth (a perception shared with governmental institutions), but to cope with this decline new, more individualized actions can be observed.

Evidence is still missing but it can probably be argued that this process of individualizing coping decisions is partly brought about by the effectiveness of national relief measures. As compared to the situation some 20 or 30 years ago, for rural households today drought is to a lesser extent a matter of life and death, and to a larger degree a problem of deteriorating assets. Consequently, many coping strategies are located within the economic sphere, and, where available, wage employment or other activities to sustain these assets become a central option. Also, the set of available strategies (partly offered by the governmental relief measures) is wider, allowing for a greater consideration of the family members' personal favours, wishes and intentions.

Adjustment and Coping Strategies Involved

With regard to the drought perceptions described above, which coping strategies are taken up by households, which are the various components of governmental relief measures, and how do they interlink?

Households have adopted a number of strategies, which can roughly be classified into three categories, i.e., autonomous (internal) adaptation measures which basically do not alter household structures (here, food sharing with other households is an important component), internal coping strategies which at least temporarily do effect household composition (e.g., looking for waged employment on the free labour market, which very often includes migration to urban centres), and finally strategies involving external assistance (especially reliance on public relief measures).

Kinship Obligations and Migration

After good rainfalls, proper grain storage and the continuous increase of the number of cattle as emergency reserves have always been part of the many autonomous adjustment strategies. If drought actually strikes and food becomes scarce, what is left is very often shared with friends, relatives and neighbours. Reciprocal kinship obligations play an important role to ensure food security within the household or village ward. Selling cattle is another strategy, and is promoted by the Botswana government, but it is usually only done when all other internal coping measures have failed. Reducing one's cattle herd means loss of wealth, but it also implies losing reputation, social prestige and influence. Animal husbandry plays a central role in *Tswana* society, and the number of cattle in one's possession is directly associated with one's social power.

One very important element of spreading risks is the out-migration of household members in order to seek employment. Remittances, either in cash or kind, sent by the migrant to his or her family in the home village may ease the recipients' financial strain brought about by drought-induced crop failure or loss of livestock. Migration of younger family members has always been a matter of negotiation and consensus on the family or lineage level (Schapera and Comaroff, 1991). Currently, however, the growing individualism concerning mediation and decision within the family has led to the point where migration may not be considered a suitable strategy for the family's well-being, but it is realized nevertheless, because it satisfies the individual preferences or needs of the migrant. In his study on food security in the village of Lethlakeng, Mazonde (1999) showed that individualization of decision making has strongly influenced and transformed drought preparedness and adaptation measures of rural households. In the many cases where family members opted for migration, and this option was based on the migrant's free will rather than on household consensus, 'implied contracts' between the remaining household and the migrant became increasingly important to ensure food-security for those who stay behind. Persons who migrate on the grounds of their individual decision also tend to determine the amount and regularity of remittances according to their own personal estimation and not so much in accordance with the actual need of the family at home. Therefore, if the family remaining in the village can secure some sort of contract with the migrant on how much money (or kind) the household should receive and how it should be used or distributed, the remittances' effect on food security becomes more calculable. In short, reciprocity of obligations has always served to increase food security. Today, with a more and more individualized decision process, to establish reciprocity becomes increasingly important, and the relevant negotiations get more complicated. The implied contracts may include looking after the migrant's affairs at home or setting up an arrangement to take care of the migrant's assets in the home village (e.g., by field ploughing or livestock herding). Mazonde (1999) states that where such implied contracts exist the level of remittances generally is higher, their frequency more regular, and their positive effect on food security more sustainable.

It must be noted that migration is not always an effective strategy. Most migrants move to urban centres to look for employment, but job opportunities are scarce and living costs in the cities high. Cash or kind transfers to rural family members therefore fluctuate and are often unpredictable and unreliable. Furthermore, the rural-urban migrant himself cannot entirely withdraw from the peril of drought. Through implied contracts, but also through the maintenance of assets many rural-urban migrants retain close links to their home villages (Krüger, 1998), but these interrelations can be disrupted or become ineffective. The rural assets, particularly cattle and access to land, usually serve as a valuable, sometimes even vital safety net for those migrants with low incomes and uncertain livelihood prospects in the city. If employment opportunities in the city subside or cash income becomes insufficient, the migrant can fall back on his or her rural assets. Because in Botswana's cities there generally is permanent access to water and food, drought normally has no direct impact on the livelihood of the migrant. The rural assets, however, are still highly susceptible to drought events and may be diminished or destroyed during drought, leaving the urban dweller without any safety buffer. Although living in an urban context migrants who have to rely on rural assets as a safety net may therefore still be put under severe threat when drought strikes.

Falling Back on External Relief

Apart from the many autonomous coping strategies, households can fall back on external relief measures. As has been outlined above, the chief objectives of governmental drought relief and recovery programmes are to insure that basic income opportunities remain available during times of drought, and to forestall an erosion of productive household assets such as arable land, breeding stock or animal draught power. There are also various direct feeding schemes, especially for nursing mothers, under five year olds and school children. The Botswana government is anxiously watching the success of its programmes and has always tried to make it clear that rural households should not rely on relief measures but attempt to adjust to and cope with drought as independently as possible. However, reliance on public schemes is probably the most important strategy of rural families today. Many households expect the government and administrative institutions to come to their assistance when there is a drought. Officials criticize that autonomous drought preparedness is constantly declining because rural families count on governmental relief should they be struck by drought-induced income losses or food shortage. On the other hand, governmental drought response has constraints and shortcomings which have contributed to the growing dependence on national relief measures.

Governmental drought relief is based on three major components that are human relief (including food aid and measures to improve the availability of drinking water and to strengthen preventive and curative health care), compensation measures for lost agricultural incomes (with a large-scale public cash-for-work scheme as a central element, but also including, e.g., subsidized purchase of livestock), and protection of productive rural assets (including livestock vaccinations and tillage subsidies).

Food Aid

The feeding programme is usually targeted at the most vulnerable groups, i.e., children and remote area dwellers. A daily school meal is provided for every pupil at primary schools. Past droughts have gone along with a rapid deterioration in the nutritional status of young children, therefore health posts now offer supplementary feeding for the under-fives. Complete cover of individual calory requirements is not intended. The meals are handed out to complement the daily household demand for food, and as a compensation for extra expenditures which arise when households have to purchase food because subsistence production has deteriorated due to drought. The feeding schemes have been successful in that the average underweight prevalence of children dropped from 30% in 1985 to an average of 13% in the 1990s (Ministry of Fincance and Development Planning, 1996). However, the feeding programmes are now beginning to lose their positive impact on the nutritional situation. Health officers had to observe that many children who obtain a daily school meal don't receive extra meals at home because their parents consider the school meals sufficient. The feeding programme therefore is a financial relief for many rural households (and thus fits in very well into the State's perception of drought causing entitlement declines), but it helped less to improve the nutritional status of children than was expected.

Employment Measures

The public cash-work-schemes form a major element of drought relief in Botswana and best reflect the State's view of drought as an entitlement problem. They are usually not targeted at specific groups, but serve as a blanket cover for all rural households who live below the poverty datum line or have little or no assets (officially, about 70% of the rural population in Botswana). In theory, most rural households affected by drought may participate in these cash-for-work schemes. It is the government's foremost aim to ensure that markets remain operational even under drought conditions. Both the demand side, i.e., rural households, and the supply side, i.e. local wholesalers and retail dealers, need support during drought. By rural works measures the government attempts to intercept financial losses and decline of household purchasing power. It is believed that with incomes earned from the participation in the programmes, households would be in a position to procure food supplies from the market (MFDP, 1992). During the 1991/92 drought, an estimated 100,000 rural households in Botswana suffered direct income losses because of the decline in domestic cereal and meat production. Through the public work schemes, which include measures to improve rural infrastructure (construction of roads, dams etc.), drought-affected people are brought in a position which allows for the puchasing of food and other goods. This, again, helps to keep local markets intact. Another aim of the employment schemes is to reduce rural-urban migration and thereby maintain a rural labour force for agricultural activities.

The large-scale public work schemes, which offer about 90,000 temporary jobs and, by a rotating system, allow the participation of up to 270,000 persons, have also been relatively successful. Free markets have indeed remained operational during the last drought periods, and income losses could be kept within limits. As yet unsolved is a problem concerning people who live in very remote areas of the country. They do not know when and where the programmes are implemented, and are not informed by officials, and therefore often cannot participate. Also, there is an increasing number of rural households who criticize the selection process at the beginning of the employment schemes as being biassed, prejudiced and discriminatory (Mazonde, 1999). Many people who are not allowed to join the programme feel they should be entitled to participate because they have suffered from drought. In fact, the criteria by which the participants are selected are unprecise, and the mechanisms involved need to be investigated and perhaps improved. The true socio-economic consequences of the employment measures for vulnerable households and village communities have not yet been clearly identified and also need further analysis. Moreover, households participating in the public cash-for-work programmes tend to neglect agricultural activities despite relatively low wages. With harvests already poor because of drought, this further increases the risk of insufficient grain yields, which again increases dependence on governmental drought relief measures. Rural-urban migration still continues, too, and there is a growing shortage of agricultural labour force in rural areas. The individualization of internal coping decisions adds to the problem. Thus, many young people opt for waged employment, either in the public works schemes or on the urban labour market, and are reluctant to take up jobs which they consider inferior, such as assisting in arable farming or herding.

Protection of Productive Assets

The measures to protect productive rural assets are probably the most questionable components of drought relief in Botswana. A major weakness here is the very bureaucratic handling of tillage subsidies. These subsidies were originally intended as an incentive for farmers to plough their land and plant their crops in time and despite uncertain rainfall conditions. The overall aims are to prevent an exodus out of arable agriculture, and to produce harvests in spite of poor rains. Payment of the subsidies is subject to the condition that a minimum of 5 hectares are ploughed at the beginning of the rainy season which normally starts in October. However, due to very formal and unsatisfactory payment procedures, most farmers don't receive their money before December or January. Therefore many farmers either don't see a point in tilling their fields earlier (i.e., in time for the crops to ripe until the end of the rainy season) or just leave their land unattended after having received the subsidies. As a result harvests are poor despite government efforts to have the fields looked after regularly. The financial burden for the State is immense: The subsidies cost the Botswana government 53.8 million Pula (around US$ 16 million) in 1994/95 alone.

Conclusions

As outlined before, there is a growing attitude among many drought-affected households to rather wait for external relief measures than to rely on traditional autonomous strategies. Also, coping with drought risk is becoming more and more a matter of individualized negotiations and decisions. The perceptions of the State and of rural households correspond in that both institutions view drought as a threat for entitlements and well-being – but there are also differences.

Drought response of rural households and the State		
	Rural households	**The State**
Overall target	Sustain livelihoods	Sustain *rural* livelihoods
Objectives and strategies aimed at mitigation of famine risk	Store food, share food, reduce food intake	Improve nutritional and health status, feed supplementary
Objectives and strategies to strengthen economic entitlements	Compensate for income losses	
	Instrumentalize reciprocal obligations	Forestall decline in household purchasing power, strengthen markets
	Withdraw from agricultural activities	Support agricultural activities
	Migrate, make use of remittances	Reduce migration
	Fall back on or sell assets	Protect rural productive assets, assist households if sales become inevitable
Other general strategies and intentions	Rely on public relief measures	Promote reliance on autonomous, traditional strategies
Objectives match	*Objectives do not entirely match*	*Objectives collide*

Figure 7.3 Drought response of rural households and the State

For the State, the major objective of drought response is to sustain rural livelihoods. For many drought-affected households, migration is a suitable response, and urban livelihoods begin to play a more important role. Some autonomous and external objectives and strategies match very well, while others contradict each other, leading to a conflict situation between what governmental institutions consider to be sensible measures, and what rural individuals or households see fit for their own well-being.

Figure 7.3 summarizes some of the perceptions, targets and response measures and illustrates where objectives meet, where they do not entirely match, and where they collide.

Although a number of intentions and actions obviously contradict each other, the conflict situations have, up to now, not endangered the basic effectiveness of drought relief in Botswana. As long as the national economy and solid financial situation of the state of Botswana permit generous drought relief measures (compared to other African countries), the programme's shortcomings may still be handled relatively easily. However, an in-depth analysis of the socio-economic impact of public schemes on rural communities and of the transformation of perception and decision processes is still pending. There is a growing uneasiness among rural households about the role governmental institutions should play. The State has tried to cut down expenses of national drought relief programmes, which amount to over 500 million Pula (almost US$ 150 million) per year if all relief measures are implemented to their full extent. Efforts to streamline the programmes are often followed by resistance from the rural communities who fear that the State wants to shift its responsibility for drought preparedness and response entirely onto them. On the other hand, a sustainable drought management which rests equally on the shoulders of both the State and the rural households, and with responsibilities for both sides made completely clear, will never be possible due to the many variations the drought process may take, and due to the constant transformation of social values and perceptions. There is no need to completely restructure drought preparedness and relief in Botswana. Despite the constraints mentioned in this paper, drought management as it presents itself today is relatively effective and successful. In general, there is a wide-spread consensus between the State and the rural communities to view drought as a threat for entitlements. This consensus forms a solid baseline for future amendments of individual, communal and governmental drought response.

References

Campbell, A. (1979), 'The 1960's drought in Botswana', in M. Hinchey and Botswana Society (eds), *Symposium on Drought in Botswana*, Botswana Society, Gaborone, pp. 98-109.

Hay, R. (1988), 'Famine Incomes and Employment: Has Botswana Anything to Teach Africa?', *World Development*, vol. 16 (9), pp. 1113-25.

Hitchcock, R. (1979), 'The Traditional Response to Drought in Botswana', in M. Hinchey and Botswana Society (eds), *Symposium on Drought in Botswana*, Botswana Society, Gaborone, pp. 91-7.

Holm, J. and Morgan, R. (1985), 'Coping with Drought in Botswana: An African Success', *The Journal of Modern African Studies*, vol. 23 (3), pp. 463-82.

International Institute for Aerial Survey and Earth Sciences (1981), *A Drought Susceptibility Pilot Survey in Northern Botswana: Final Report*, ITC, Enschede.

Krüger, F. (1997), *Urbanisierung und Verwundbarkeit in Botswana*, Sozio-ökonomische Prozesse in Asien und Afrika No. 1, Centaurus, Pfaffenweiler.

Krüger, F. (1998), 'Taking Advantage or Rural Assets as a Coping Strategy for the Urban Poor', *Environment and Urbanization*, vol. 10 (1), pp. 119-34.

Mazonde, I. (1999), 'Social Transformation and Food Security in the Household: The Experience of Rural Botswana', in F. Krüger, G. Rakelmann and P. Schierholz (eds), *Alltagswelten im Umbruch*, LIT, Münster, Hamburg, London (in press).

Ministry of Finance and Development Planning (1991), *National Development Plan 7*, MFDP, Gaborone.

Ministry of Finance and Development Planning (1992), *Aide Memoire: The Drought Situation in Botswana and the Government Response*, MFDP, Gaborone.

Ministry of Finance and Development Planning (1996), *The National Nutrition Surveillance System*, MFDP, Gaborone.

Mokobi, K. and Asefa, S. (1988), 'The Role of the Government of Botswana in Increasing Rural and Urban Access to Food', in M. Rukuni and R. Bernsten (eds), *Southern Africa: Food Security Options*, Michigan State University, University of Zimbabwe, Harare, pp. 257-73.

Republic of Botswana (1985), *Report on the National Food Strategy*, Rural Development Unit, Gaborone.

Republic of Botswana (1995), *National Food Strategy*, Republic of Botswana, Gaborone.

Sandford, S. (1979), 'Towards a Definition of Drought', in M. Hinchey and Botswana Society (eds), *Symposium on Drought in Botswana*, Botswana Society, Gaborone, pp. 33-40.

Schapera, I. and Comaroff, J. (1991), *The Tswana*, Kegan Paul International, London.

Sen, A. (1981), *Poverty and Famines: An Essay on Entitlements and Deprivation*. Oxford University Press, Oxford.

Teklu, T. (1994): 'The Prevention and Mitigation of Famine: Policy Lessons from Botswana and Sudan', *Disasters*, vol. 18 (1), pp. 35-47.

8 Environmental Conservation, Land Tenure and Migration

The case of the Atlantic rainforest in southeast Brazil

FLORIAN DÜNCKMANN

Introduction

Recent developments in the Amazon region are a central matter of concern in the current discussion on environmental conservation in neo-tropical rainforests. However, far less attention has been dedicated to the Atlantic Rainforest. It is much smaller in size but nonetheless needs effective conservation measures to protect its enormous biodiversity. What is especially worrying is that not only 10% of its original stands is left (Câmara, 1991; Dean, 1996). The Atlantic Rainforest is considered to be the most endangered rainforest in the western hemisphere (Dinerstein, Olson, Graham, Webster, Primm, Bookbinder and Ledec, 1995).

In its original form, Atlantic rainforest vegetation covered almost all slopes and mountain regions along the Brazilian Atlantic coast, i.e., from the so-called *zona da mata* in the northeast to the State of Rio Grande do Sul in the south. Following Portuguese colonization, massive usages came along with forest destruction, namely by the extraction of Brazil Wood (*Caesalpina echinata*) and by cultivating sugar cane. At the end of the last century, with Brazil becoming an independent state, the coffee boom set in, especially in the southeastern states of Rio de Janeiro, Minas Gerais and São Paulo. With the emergence of coffee being the region's leading economic activity in a national context plantations were expanded into the formerly sparsely populated woodlands of the São Paulo hinterland.

Vale do Ribeira: Regional Setting and Environmental Conservation

In the wave of coffee plantations to be established on a large scale, the humid and warm coastal areas in the Vale do Riberia of Southeast Brazil, however, turned out to be not suitable at all for cultivation. Thus, the region to be explored here in more detail has acquired a special status within São Paulo state's economy that is exemplified as follows.

With regard to the economic structure, the region is dominated by primary sector activities. Being characterized by low average incomes and a significantly higher share of families living below the line of poverty (Dünckmann, 1996), Vale do Ribeira constitutes a 'pocket of poverty' as compared to the otherwise economically more dynamic regions of the state. The main focus of agriculture in the region is on capital-extensive forms of production, namely the cultivation of bananas, rice, maize and beans in small-scale subsistence farming. As compared to this, farming in the rest of the state of São Paulo has recently been characterized by Silva (1993) in terms of modernization and industrialization.

In Vale do Ribeira, the issue of land tenure is not yet settled. The majority of smallholders do not have an official title on their land and have to rely on informal land rights. This applies to the traditional groups of population that have been farming the area for generations, but also to recently arrived families. A parallel market with informal and fraud land titles as well as legal and official land use concessions has emerged and can hardly be differentiated from formal real estate trade. Due to this legal insecurity in terms of land tenure, land conflicts are frequent in the region even causing a number of deaths in the recent past (Brito, 1987; Perosa, 1992). As a result, large-scale land clearances and felling of trees are done just to demonstrate the claims of owners giving no further and productive uses to the land acquired.

Since a 'victory of coffee cultivation' did not occur in Vale do Ribeira, causing the destruction of natural forests elsewhere in São Paulo state, vast stands of the Atlantic Rainforest still exist especially on the steep slopes of Serra do Mar. In the area, further types of endangered ecosystems such as mangroves and dune forests could be found, too.

Due to its prominent position in terms of vegetation, Vale do Ribeira became the focus of official efforts aimed at protecting the Atlantic Rainforest. As depicted in Figure 8.1, more than 50% of the state's strict nature reserves are concentrated in the area (Capobianco 1994), and more than 25% of the region's surface falls under juridical categories according to which any economic activity is prohibited. The nature reserves include *Parque Estadual* and *Estação Ecológica* being classified in terms of category II (National Park) and, respectively, category I (Scientific Reserve) under the classification system brought forward by the International Union for Conservation of Nature (IUCN). While the establishment of protected areas was first reinforced in the 1980s, the Ministry of Environmental Conservation (Secretaria do Meio Ambiente, SMA) plans to continue with the expansion of protected areas in the near future (SMA, 1992).

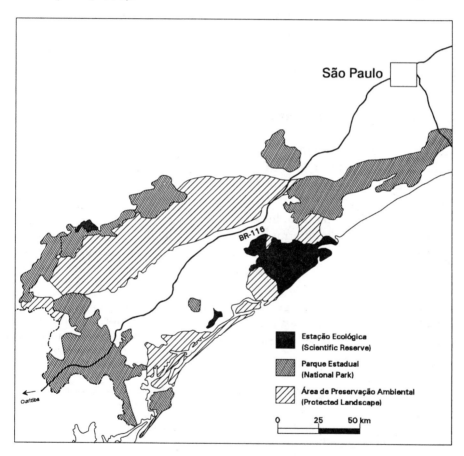

Figure 8.1   Protected areas in the Vale do Ribeira

In the realization of environmental protection measures, many problems become obvious. Although it is officially prohibited, almost all reserves are inhabited and economically used. Many families settled in these areas before they were declared nature reserves and now find themselves as illegal residents in strictly protected areas. Other families settled there after the official declaration. This was possible because there is no effective control over the vast areas due to an understaffed park administration. Even today the residents' legal status has not been settled although many reserves were implemented more than 10 years ago. Thus neither local residents nor reserve authorities know who may remain resident in the reserves, what kind of restrictions in terms of economic activities would be effective, and what is to be done with the rest of the settlers (Wehrhahn, 1994). Obviously, the unclear situation has caused tensions among local residents and nature reserve administration. In the following case study, conflicts as well as unintended social consequences arising from nature reserve policy are presented.

Parque Estadual Jacupiranga: A 'Paper Park'

Parque Estadual Jacupiranga is one of the oldest and most conflict-ridden nature reserves in the region. The first initiative to protect the area was taken as early as in 1945. It was then declared a *Reserva Florestal* though without any clearly defined legal status. In 1969, it officially became Parque Estadual Jacupiranga. In the area totalling 150,000 hectares of land, any economic activity in terms of land use (agriculture or extractivism) is prohibited by law with 99% of all land falling under strictly to be conserved natural ecosystems. In addition, it has to be stressed that the total of the area was supposed to be state-owned.

However, real conditions look somehow different if not giving a completely divergent picture as compared to the situation laid down on paper. Only five years ago, reserve authorities actually started to turn the 'paper park' into reality (Fundação SOS Mata Atlantica, 1993). Implementation turned out to be very difficult due to the too long lack of activity from state authorities. Today, an estimated 2,000 families live in the area. Land tenure is not settled due to the above mentioned practice of squatting. Transforming the area into state-owned land would be a very long, complicated and costly process for which hardly any funds could be raised.

What is far more problematic than the legal situation, however, is the actual intense land use on reserve grounds – see Figure 8.2. Particularly in the western and eastern fringes, a massive expansion of extensive cattle farming onto reserve ground could be identified. Along the federal highway BR 116, connecting the city of São Paulo with the southern states (as well as with Uruguay and Argentina), illegal occupation by new settlers is common practice. According to estimations done by SMA (1992), as many as half of the nature reserve's population came to settle along the highway built in the 1960s.

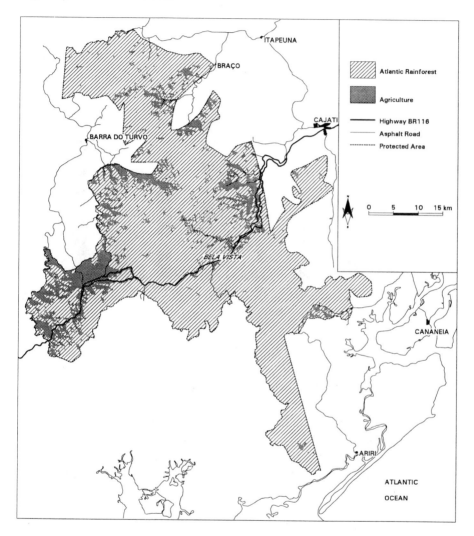

Figure 8.2   Land use in the Parque Estadual Jacupiranga
Source: Adapted from Fundação SOS Mata Atlântica (1993)

The Case of Bela Vista Settlement

Situated at BR 116, Bela Vista is one of the new settlements along the highway. The following findings on the settlement's specific situation are based upon interviews to get a picture of socioeconomic features (e.g., demographic structure, livelihood strategies and history of migration) and the population's views on nature reserve policy. The interviews were conducted in 48 households of Bela Vista having a total of about 100 families that migrated to the place during the last 5 years – see Figure 8.3.

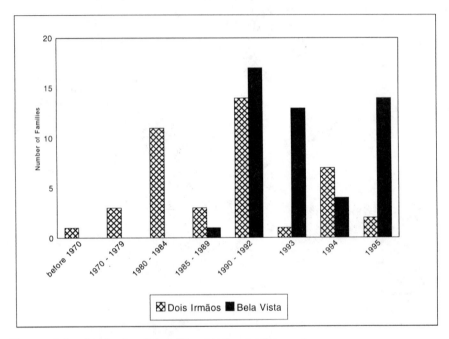

Figure 8.3   Arrivals of families in the settlements

The settlers are predominantly of the second generation. The majority bought their land from the original *posseiros* (squatters without title), i.e., not receiving it under official titles but holding it informally. According to the interviewees, they nevertheless had to pay a sum of money for their moderate property for the land.

Preceding their arrival at Bela Vista, and except for a single family, all other settlers used to live in the city of Curitiba being the capital of Paraná at a distance of 150 km from Bela Vista. Apparently, it was urban-rural migration that created the settlement what is very unusual in the context of a massive rural exodus as commonly to be found in Brazil. However, a closer look at the history of migration provides a more differentiated picture.

All household heads interviewed used to spend their childhood and youth in the countryside with the majority of them even having spent almost all their lives there. Many interviewees reported that they did not get used to city life and that their most important motive to migrate was resuming agriculture. Furthermore, increasing unemployment in the cities functioned as a push-factor of migration. Although rural unemployment was perceived as a threat, too, people considered it to be more favourable if they could provide for their basic needs without having to pay high rents. One dominant economic motive for migration was reducing the costs of living instead of finding new sources of income.

Among the persons interviewed, 42 out of 48 household heads had in common that they used to live in rural northwest Paraná before moving to Bela Vista. They had been smallholders there or used to work as seasonal or day laborers on large farms. By then, northwest Paraná had been the dynamic pioneering frontier up to the 1960s being characterized by considerable immigration (Kohlhepp, 1975). In the 1970s, however, the modernization of agriculture and changing production structures caused the dissolution of traditional labor structures and created a large number of redundant workers. According to Coy (1988), rural Paraná became a 'consolidated pioneering frontier' providing many emigrants who were made redundant and marginalized in the course of the modernization of agricultural production. The emergence of ever new pioneering frontiers (such as Amazonia) functions as an outlet for social conflicts which would otherwise inevitably shake the rural areas affected. In fact, after having left Paraná, 7 out of all household heads interviewed migrated to other recent pioneering frontiers in Amazonia or Paraguay before arriving at the city of Curitiba and Bela Vista, finally.

Given the fact that smallholders are threatened by the expansion of modern agriculture and that areas where landless families could settle are virtually non-existent, it could be stated that nature reserves act as a pull-factor leading to uncontrolled squatting in Bela Vista area. Many interviewees said that they were indeed surprised to find land in such a location that was not being used. On the one hand, and though the area is a nature reserve where any settlement was supposed to be strictly prohibited by law, squatting nevertheless occurred, while on the other hand squatting happened since the area precisely was a nature reserve that provided abundant room for occupation.

Conflicts with the Nature Reserve Administration

Settlers emphasize that they did not know of the existence of the nature reserve when they bought the land. However, the authorities insist on their officially illegal status resulting from the lack of land titles. Therefore, in case of expropriation, settlers would only be entitled to compensations for all immobile goods except for land, i.e., their cheaply built huts. So far, however, conservation authorities have not yet started an initiative to handle the insecure land tenure situation. Though, at present, administration blocks the provision of infrastructure such as access to roads and electricity and makes Bela Vista a place that can only be reached through a dirt road unsuitable for cars and hardly to be used during rains. On the one hand, park administration fears to enhance further immigration and illegal commercial clearcut if they provide infrastructure. On the other hand the park police cannot adequately patrol the area due to the lack of infrastructure. This explains why only three of the interviewees stated to have ever met any conservation authority representative. Thus agriculture, although practised within a strictly protected nature reserve, can be carried out virtually unrestricted.

The Soil Mining Character of Agriculture

It could be doubted whether agriculture will establish as the most important economic activity in the long run since it has by no means proved to be sustainable up to now. Primary and secondary forests were cut in order to create fields on which subsistence products such as maize, rice, beans or manioc are grown. Due to the short period of settlement, no stable rotation system for farming has been established. However, and considering the relatively small size of the allotments (i.e., 10 to 20 ha per family), rotational farming is not very likely put into practice. Sustainable land use either requires inputs of fertilizers (a not very realistic option due to the lack of capital found) or large allotments that allow for the inclusion of adequate fallow periods. Presently, land managers can still profit from natural nutrients provided from soils and vegetation by slashing and burning the well developed forests. Unless sustainable forms of farming are established, people will have to leave the area as soon as the remaining forests will have been cleared. Even without the driver of population growth, there is any reason to assume that this is likely to happen in the near future. Southgate and Clarke (1993) characterized the current system of extractive land use consuming ever new forests to constitute the case of 'nutrient mining'.

Problems Associated with Protected Areas

Establishment and Management of protected areas, though seemingly not related to social processes at first glance, prove to be well affected by but also inducing social processes from which they simply cannot be isolated. The issue of land tenure in the whole of Brazil shows direct linkages to the destruction of tropical rainforests since they constitute remaining areas offering space for expansion particularly for landless farmers. What also becomes clear in the case of Bela Vista is the problem of how to design an adequate concept of conservation if wilderness areas are to be protected. Considering the large size of the reserve, efficient controls and patrols are not feasible, especially if conservation measures are to be applied in purely restrictive terms, thus, causing conflicts with local populations.

However, first steps are taken in Brazil to implement integrative strategies in order to reconcile the use of land with the protection of biodiversity, i.e., to overcome the dichotomy of nature conservation and destructive usages. Nevertheless, even the new measures designed are confronted with social problems as it is illustrated in the following case.

Sectoral Conservation Regulations

In the 1980s, the protection of the Atlantic Rainforest became a major political goal on the national as well as on the statal level. In the Constitution of 1988, the Atlantic Rainforest – along with other types of ecosystems such as the Amazon forests – was defined in terms of a 'national heritage' (Art. 225, § 4). And with the national decret 750 (*Decreto no.* 750/93) the status of protection as well as the restrictions in terms of usages were specified (Art. 1) in that slashing and exploring primary or medium and advanced stages of regeneration are generally prohibited. Only with the consent of local conservation authorities, initial clearances for agricultural purposes may be performed.

As mentioned above, the overwhelming majority of land users in Vale do Ribeira does not have official land titles but more so relies upon informal common law. Most of the residents are also illiterate and only dispose of scarce money. Thus, administrative steps needed to be taken to apply for an official land use permission constitute virtually unsurmountable problems to be (un)solved by nearly all of the settlers. In addition, local conservation authorities are inadequately provided with money and staff so that applications easily take a couple of years to be processed. In summary, the *Decreto No. 750/93* in reality prohibits any use of initial stages of regeneration in Vale do Ribeira. The consequences of these restrictions for agriculture as well as for related social processes in region of Vale do Ribeira will be dealt with in more details as follows. The findings are based upon interviews conducted in 50 households of Dois Irmãos settlement.

The Case of Dois Irmãos

As could be derived from Figure 8.3, and different from the case of Bela Vista, in the settlement of Dois Irmãos more of recently immigrated families but also families who arrived here a longer time ago could be found. The major difference is good infrastructure, in particular road access through a relatively well fixed dirt road. The settlement is also not located in a nature reserve. Due to its easy access, the forest police frequently patrols in Dois Irmãos. Accordingly, interviews proved that the direct impact of conservation policy on land use is far stronger than in Bela Vista, which was a puzzling result at first sight. About one third of the interviewees were fined once or more for violations of conservation regulations. The most frequent offences were clearing secondary vegetation and slash-burning remaining vegetation, i.e., activities which normally go hand in hand with shifting cultivation.

What needs to be taken into account here is that restrictions mainly affect annual cultures such as subsistence products. Due to a lack of money farmers cannot afford mineral fertilizers, and farming, thus, demands traditional rotation systems in order to allow adequate soil regeneration. Shifting cultivation implies clearing areas with relatively well developed secondary vegetation. Due to the capital-extensive strategy of land use, farmers conflict with the goal of conserving nature. More capital-intensive varieties of farming, such as pasture-land use or cultivating bananas cause fewer conflicts because areas are constantly used for a couple of years instead of single year usages as in the case of shifting cultivation.

Effects of Conservation Regulations

Agriculture turns out to be the main income source for only 25% of the households interviewed. In general, agriculturalists will lay out their fields in proper distance from the road with forest police officers constantly patrolling there. Another important strategy of subsistence for other settlers than pure agriculturalists is work on banana plantations as daily laborers. Families still using farmland often transformed it into pasture since this type of farming was not seen to conflict with conservation regulations. Other strategies of subsistence include sales of land, migration to the small town of Sete Barras, and working as caseiros on farms owned by people from the city of São Paulo. In the latter case, those farms are only used for weekend leisure or holidays by the city based owners with about one fourth of them run by caretakers who generally are paid the minimum wage. This type of agriculture, i.e., mostly cattle and horse ranching, hardly conflicts with conservation regulations since own cropping only plays a minor role in the household strategies of settlers involved there.

Thus, measures of environmental protection contribute to the decrease of small-scale farming and reinforce a social process that has sustainably influenced large parts of Vale do Ribeira since the 1980s. Namely, more and more autonomous smallholders become dependent laborers or stopped farming own land by large. It was noted by Bianchi (1988) that autonomous agriculture, which used to be the dominant form of economic activities in the area, has now become one out of many other strategies of subsistence. The process of substituting smallholders by leisure farms financed by urban dwellers is seen to be on an increase since the traditional mode of farming has become restricted by regulations aimed at conserving wilderness.

Conclusions

Though widely different from each other, the case studies presented here from the southeast of Brazil prove that conservation measures well affect social processes and could have (unintended) consequences running counter to the original intentions of conservation policy.

Strict nature reserves, i.e., a complete separation of 'nature' and use, as well as sectoral conservation regulations including the implementation of conservation goals in areas used for agriculture turn out to be highly problematic. Full area protection against settlement and deforestation is hardly feasible without active and effective participation of the local population, especially if there are urgent needs for land to survive on. Sectoral regulations which in theory should dissolute the dichotomy of conservation and land use are often inadequate with regard to the local situation and can thus only restrictively impact upon the usage of land.

Considering the strong social polarization among poor and rich in rural Brazil, it needs to be mentioned that conservation regulations affect rich and poor people to a highly various extent. Particularly low-income families being directly dependent on the use of natural resources for subsistence will have to carry the main and larger burden. On the other hand, the case of Bela Vista has shown that 'shifted cultivators' (Myers 1993), i.e., marginalized, landless families on the search for land, could be easily perceived as impacting negatively upon remaining stands of natural forests.

Thus, the warnings as expressed by Ghimire (1991) that nature conservation may be a major source of future conflicts arising in developing countries needs to be taken seriously as could be seen from the case of Vale do Ribeira. If conservation concepts will only be restrictive and implemented by a top down process without respecting the forms of land use of the local population and without offering sustainable alternatives, the remaining stands of the Atlantic Rainforest cannot be protected in the long run, but only survive temporarily 'in close custody'.

References

Bianchi, A.M. (1988), *Mobilidade: Estratêgia de Sobreviver*, Editora Instituto de Pesquisas Econômicas, Universidade de Sao Paulo.

Brito, M.C.W. (1987), *The Land Conflict Resolution Team: An Approach to Implementation Problems in São Paulo-Brasil*, Ottawa.

Câmara, J.G. (1991), *Plano de Ação para a Mata Atlântica*, Fundação SOS Mata Atlântica, São Paulo.

Capobianco, J.P. (1994), 'O Vale do Ribeira É um Desafio', in Konrad-Adenauer Stiftung and Fundação SOS Mata Atlântica (eds), *Mata Atlântica e a Imprensa: Relatório do Laboratório Ambiental, Vale do Ribeira*, Fundação SOS Mata Atlântica, KAS, São Paulo, pp. 9-11.

Coy, M. (1988), *Regionalentwicklung und regionale Entwicklungsplanung an der Peripherie in Amazonien: Probleme und Interessenkonflikte bei der Erschließung einer jungen Pionierfront am Beispiel des brasilianischen Bundesstaates Rondônia*, Tübinger Geographische Studien No. 97, Geographisches Institut der Universität Tübingen.

Dean, W. (1996), *A Ferro e Fogo, A História e a Devastação da Mata Atlântica*, Companhia das Letras, São Paulo.

Dinerstein, E., Olson, D.M., Graham, D.J., Webster, A.L., Primm, S.A., Bookbinder, M.P. and Ledec, G. (1995), *A Conservation Assessment of the Terrestrial Ecoregions of Latin America and the Carribean*, World Bank, Washington, DC.

Dünckmann, F. (1996), 'Vale do Ribeira, Brasilien: Agrarstruktur einer zentrumsnahen Peripherie', in P. Gans (ed.), *Regionale Entwicklung in Lateinamerika*, Erfurter Geographische Studien No. 5, Geographisches Institut der Universität Erfurt, pp. 189-208.

Fundação SOS Mata Atlântica (1993), *Parque Estadual de Jacupiranga: Diagnóstico Preliminar*, Fundação Mata Atlântica, São Paulo.

Ghimire, K.B. (1991), *Parks and People: Livelihood Issues in National Parks Management in Thailand and Madagascar*, UNRISD Discussion Paper No. 29, United Nations Research Institute for Social Development, Geneva.

Kohlhepp, G. (1975), *Agrarkolonisation in Nord-Paraná: Wirtschafts- und sozialgeographische Entwicklungsprozesse einer randtropischen Pionierzone Brasiliens unter dem Einfluß des Kaffeeanbaus*, Heidelberger Geographische Arbeiten No. 41, Steiner, Wiesbaden.

Myers, N. (1993), 'Tropical Forests: The Main Deforestation Fronts', *Environmental Conservation*, vol. 20 (1), pp. 9-16.

Perosa, E.P. (1992), *A Questão Possessoria no Vale do Ribeira – São Paulo: Conflito, Permanência e Transformação,* Trabalho de Mestrado, Universidade de São Paulo.

Silva, J.G. (1993), 'A industrialização e a Urbanização da Agricultura', in Fundação Sistema Estadual de Análise de Dados (ed) *O Agrário Paulista,* SEADE, São Paulo, pp. 2-10.

Secretaria do Meio Ambiente, Estado de São Paulo (1992), *Plano de Ação Emergencial. Implantação e Manejo de Unidades de Conservação,* Secretaria do Meio Ambiente, São Paulo.

Southgate, D. and Clark, H.L. (1993), 'Can Conservation Projects Save Biodiversity in South America?', *Ambio,* vol. 22 (1), pp. 163-66.

Wehrhahn, R. (1994), *Konflikte zwischen Naturschutz und Entwicklung im Bereich des Atlantischen Küstenregenwaldes im Bundesstaat São Paulo, Brasilien: Untersuchungen zur Wahrnehmung von Umweltproblemen und zur Umsetzung von Schutzkonzepten,* Kieler Geographische Schriften No. 89, Geographisches Institut, University of Kiel.

9 Why Herd Animals Die
Environmental perception and cultural risk management in the Andes

BARBARA GÖBEL

Introduction

In academic and popular debates about the impacts of global climatic change arid ecosystems, together with tropical rain forests, are presented as paradigmatic examples of fragile and endangered ecosystems, which urgently require protection. Following the general debate on sustainable development, which has been furthered by Agenda 21 of the United Nations Conference in Rio de Janeiro (1992) and by the UN declaration of the Decade of the World's Indigenous People, a broad consensus exists in that decision-making concerning the long-term protection of an endangered ecosystem must involve the indigenous inhabitants of the area. For overviews on these developments see, for example, Antweiler (1998), Escobar (1995), Milton (1996), and Orlove and Brush (1996).

However, the design and realization of concrete measures frequently leads to conflicts due to different ideas concerning the environment, irreconcilable practices in the use of resources, or unequal power structures. Collisions are commonly found to occur between environmental goals aimed at by western industrialized countries, the economic and political interests of national and regional power groups, and the local economic practices and knowledge systems. The question as to what exactly the concrete risks for an arid ecosystem are, and to what threats the inhabitants of such an ecosystem are exposed to, can therefore be answered quite differently depending on the point of view and position of the agents involved. A certain degree of agreement on this question is, however, a necessary precondition for sustainable development which takes account not only of ecological, but also of economic, social and cultural factors. One way to make such consensus possible is to appreciate cultural differences in the perception of dangers and the management of risks – to a greater extent than has generally been the case in the debate on endangered ecosystems. It would also be important to try and understand more exactly which other conceptions of nature these differences are related to. The present contribution, which focusses on the links between environmental perception and cultural risk management in the small Andean community of Huancar (north-western Argentina), is intended as a small step in this direction.

The Pastoral Community of Huancar

Huancar[1] is situated on the eastern edge of the Puna de Atacama, which constitutes one of the most arid sectors of the Andean highlands (Troll, 1968). The area is characterized by the prevalence of salt lakes, volcanoes, and scattered grass and shrub vegetation. In political and administrative terms the District of Huancar (*Distrito Huancar*) forms a part of the Department of Susques (*Departamento Susques*) in the northern Argentine Province of Jujuy – see Figure 9.1. Because of its aridity and its inacessibility the region has always been at the periphery of the provincial and the national states (Delgado and Göbel, 1995). Only in the last few years this situation has changed slightly. Because of the inauguration of a transnational road running to Chile, an exceptionally violent eruption of a volcano, and a dramatic drought cycle, the centers of political power (i.e., mainly the provincial capital Jujuy and the national capital Buenos Aires) started to take more notice of the Department of Susques. The Department developed into a potential target for national and international aid funds. The mediation of such funds constitutes an important power instrument for the state, the political parties and the Catholic Church since it provides them with the opportunity of building up and reinforcing clientelistic networks in remote rural areas. From 1996 onwards, a hitherto unknown activism on the part of government, parties and ecclesiastical institutions broke out in the whole Department of Susques.

The District of Huancar covers an area of about 20 km by 27 km, being the extension of the whole Department of Susques 9,199 km². Though largest in size among all the province's departments, the Department of Susques is characterized by the lowest density of population, i.e., 0.3 persons/km² (Quinteros and Masuelli, 1991). According to official census data from 1991, it has a population of 2,847 persons, living 671 of them in Susques, which is a small town and the administrative center of the area. The same census reports a total of 513,992 inhabitants for the province of Jujuy (53,219 km²) and a population density of 9.7 persons/km². While census data of 1991 do not allow for a breakdown of population by the

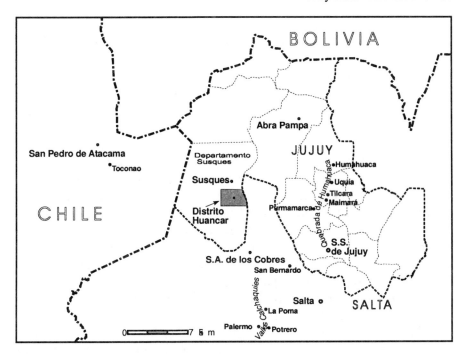

Figure 9.1 Location of the Distrito Huancar

department's districts (*distritos*), in the case of Huancar the district's population reportedly increased from 310 inhabitants in 1993 to 330 in 1996.

In contrast to historical descriptions from the 18th and 19th centuries, the present inhabitants of the area do not refer to themselves as members of an autonomous ethnic group and no longer speak the former indigenous languages of the Puna de Atacama (*Kunza* and *Quichua*). But many of their practices and ideas make them culturally different from the inhabitants of wealthier sectors of the highlands of Jujuy, of the agricultural valleys, and of course from those in the cities. Most of their specific cultural features, as people in Huancar say themselves, are related to the mobile herding of llamas, sheep and goats. For the majority of the inhabitants animal husbandry is not only of great economic importance but also constitutes a central element in their cosmology (Göbel, 1998 a).

In 1993, 38 of the 46 households in the District defined themselves as pastoral households; three years later the proportion was 36 out of 45 households.[2] The average herd of these pastoral families comprises 178 animals, of which 26% are llamas, 36% goats and 38% sheep. However, considerable differences exist between households with respect to herd size. The number of animals can range from around 50 animals (as found in 4 households) to 350 animals (3 housholds) or even 613 animals (1 household). In addition, each household possesses at least half a dozen donkeys. Households that often organize trade caravans own 50 or more donkeys and even one or two mules, which are important male prestige objects. In contrast to sheep, goats and llamas, the donkeys and mules are not herded and move around freely.

Those families in Huancar that do not engage any longer in pastoralism (8 families in 1993, 9 in 1996) still own some llamas, sheep or goats.[3] They give these animals to other households for herding, maintaining them as 'savings just in case'. While these non-pastoral families mainly stay in the small hamlet of Huancar situated in the center of the district, pastoral life is oriented towards the mountains. There the herders' central farmsteads, pasture grounds and herding stations are located. They are scattered in an harsh environment at an altitude of 3,600 m to 4,300 m with an average pasture ground sizing 1,649 ha.

In contrast to other parts of the Andes (Brush and Guillet, 1985; Browman, 1974; Orlove, 1981; West, 1983), no communal pastures or water-holes exist in Huancar. Each family holds exclusive rights to certain pasture grounds and waterholes. However, concerning the gathering of firewood or the grazing of donkeys and mules, there are no territorial restrictions. Another specific feature of this part of the Andean highlands is the high mobility of the herders with their llamas, sheep and goats (Flores Ochoa, 1968; Orlove, 1981; Palacios Ríos, 1981; West, 1983). Approximately every three weeks families move with their herds from one of their many herding stations to another or back to the farmsteads. This high mobility of the herds as well as the isolation of most farmsteads probably contribute to the conceptualization of households as autonomous social and economic units. They are considered to be the central arena of production, consumption and social reproduction in Huancar society.

Within the pastoral economy, at least on a normative level, there is a marked division of labor by gender. It is characterized by the contrast between the local orientation of women and the extensive mobility of men (Göbel, 1997 a, 1998 a). The daily herding and management of the flocks are considered to be female work. Women are assisted in this by their children, especially their daughters and granddaughters, when they do not have to attend school. The support of adult men is only necessary for those pastoral activities which require either great physical strength (e.g., dipping, or building and repairing of enclosures) or special male skills (e.g., castration). However, these exceptional interventions by men in animal husbandry are not socially considered to be a prominent male domain, which is rather the trading of local products through trade caravans with donkeys and mules (Göbel, 1998 b, 1998 c). In the pastoral economy of Huancar economic exchange is of eminent importance. Cultivation is practically impossible in this part of the highlands. Agricultural products such as maize, potatoes and beans, which are basic components of the daily diet, therefore have to be acquired from lower regions. To a great extent they are bartered for highland products such as woven goods, knitted ware, cheese, hides, dried meat, and salt from the nearby salt lake. The principal exchange circuits of the trade caravans include the valleys in the east (Quebrada de Humahuaca) and south (Valles Calchaquíes) of the District, as well as the Chilean oasis near San Pedro de Atacama – see Figure 9.1. The other main area of male competence is supplementing pastoral production through temporary wage labor in mines or in tobacco and fruit farms in the lower, subtropical parts of north-western Argentina.

Risks for Pastoral Production: *problemas* and *peligros*

Pastoral households in Huancar have to deal with a set of problems which are related to their unpredictable living conditions. These problems can be characterized in an abstract manner by having recourse to the framework of risk analysis. They can have different repercussions on the households. Variations are related to differences in household structure, the size and composition of herds and pasture grounds and the access to income alternatives such as temporary jobs (while intra-household variability of risks impacts are not considered here).

People in Huancar do not use the term 'risk' when they refer to the threats to their economy, but speak of *peligros* (dangers) or *problemas* (problems). They describe *peligros* or *problemas* as events or situations having a negative impact, which may occur at any time and with any intensity, and against which very little, if anything, can be done. Of central importance in their charcterizations of *peligros* or *problemas* are the aspects of unpredictability, endangering and loss (see also the indigenous risk evaluation criteria described below). These terms correspond to the core elements of common scientific definitions of 'risk' (Fischhoff, Lichtenstein, Slovic, Derby and Keeney, 1981; Kasperson and Kasperson, 1987; Renn, 1998; Yates and Stone, 1992), so that it seems justifiable to use the concept for a general analysis of herd management in Huancar.[4] For the sake of simplicity, I will therefore use the terms 'risks', 'problems' and 'dangers' as synonyms in this article. But I want to emphasize that this proceeding can only represent a first, preliminar approach, since it implies a mixture of different explanatory perspectives: the point of view of the pastoralists that are directly affected by the risks and the more distant analytical perspective of researchers. It should not be allowed to hide the fact that risk perception and risk management are cultural phenomena and can therefore differ from culture to culture.

Moreover, for reasons of space I shall limit myself here to the analysis of two central fields of risk for the pastoral economy in Huancar. One of these problem areas can be summed up by the general term 'ecological risks'. It constitutes the classic focus of ethnological risk studies (Barlett, 1980; Blaikie, Cannon, Davis and Wisner, 1994; Cashdan, 1990; DeGarine and Harrison, 1988; Halstead and O'Shea, 1989). The other problem area can be expressed by the term 'institutional risks'and has not been treated very extensively up to now. With regard to pastoralism, see, for example, Bollig (1997), Fratkin (1997), Fukui and Markakis (1994) and Göbel (1997 b).

Ecological Risks

Because of the difficult ecological conditions in this extremely arid sector of the Andean highlands, the study area constitutes a paradigmatic case for analysing the impact and management of a risky environment. Annual rainfall is very low in the area. It is restricted to the southern summer (December to March) and in general does not surpass 200 mm (Bianchi, 1991). Moreover, precipitation is subject to great stochastic variations. According to oral traditions, it either does not rain at all in three out of ten years or rains last only on a few days. Also the occurrence of frost, hail, snow and strong winds fluctuates enormously from one year to the next. Although herders dispose of long-term experience of the periods within which specific climatic events are concentrated, they cannot predict the actual occurrence and scale of the event. Related to climatic fluctuations, the size and composition of available pastures and waterholes vary stochastically from year to year. The variations in the impact of predators such as pumas and foxes seem to be connected to these climatic conditions, too.

Two consequences of these environmental risks are that the health of the herd animals and herd growth are subject to great fluctuations. In average years 30 to 50% of new-born and young animals die, while in bad years the losses may reach 80% or more. A woman with a herd of about 300 animals illustrated this as follows: 'In a normal year from 80 ewes we get 50 lambs. Of these lambs only 25 to 30 survive the first year of life'. She mentioned the same numbers for 70 female goats. With regard to llamas she said: 'In an average year 50 llama mares will produce 25 foals. But from these foals only 12 to 15 *tequis,* llama-foals, will survive.' Although the mortality of adult animals is not as high as the mortality of new-born or young animals, it is also perceived to be a severe threat to a pastoral household. It implies the loss of considerably more meat, wool and labour investments than the loss of a new-born or young animal imply. The most important consequence of the great potential danger of animal losses is that herders cannot reliably predict what quantities of meat, wool, hides, milk and cheese they will be able to dispose of. These production uncertainties not only affect decision-making concerning herd management (e.g., frequency of spatial movements with animals, fodder and work force requirements), but also decisions on consumption patterns and reliance on other income possibilities. They also have a severe impact on economic exchange. In view of these imponderabilities long-term economic planning is impossible.

The problem variables briefly outlined here which result from my general analysis of pastoralism in Huancar also turn up in the day-to-day conversation of the herders or were specified as such when responding to my direct questions.[5] Thus the herders consider drought (*sequía*), the death of a herd animal in general (*muerte de hacienda*), lightning (*rayo*), hail (*granizo*) and snow (*nieve*) as central problems and dangers for life in the Andean highlands. However, they do not aggregate risks by groups. For example, they do not think of all climatic events as a single group. They do not speak in a general way of 'environmental problems or dangers'. It would therefore seem that the way the highland dwellers understand their environment differs from western scientific ideas.

The Andean evaluation of the risks mentioned above is firmly grounded in practice. Risks are classified by the herders according to three non-exclusive criteria. I want to present these criteria by reproducing the indigenous classification of dangers for a herd animal. One criterion for risk evaluation is the degree of exposure to a danger. For example, a distinction is made according to whether a risk affects only one or a few animals (as in the case of lightning or being killed by a puma or fox) or whether all the animals are affected (as in the case of a drought). Another criterion is the impact of a negative event. Herders distinguish between risks which are highly likely to have fatal results (e.g., lightning, puma or fox) and risks which may only cause an animal to fall sick. The third criterion is the extent to which a risk is controllable. Here a distinction is made between risks with a known cause, where direct measures can be taken in order to minimize damages and those, where the cause is unknown. Ritual practices (e.g., sacrifices, prayers), as we shall be seeing, are important as preventive measures for all kinds of risks. But while in the case of hail showers it is possible to reduce the extent of the damage by keeping the animals strictly under control, in the case of drought no direct countermeasures are possible.

Institutional Risks

The other central problem for the pastoral economy in Huancar treated here is institutional risks. Pastoralists address this problem with the general concept of gobierno. They consider gobierno to be as dangerous as a drought and say that these two are the greatest problems for pastoral life. However, for the inhabitants of this remote part of the Andean highlands gobierno does not only mean the actual departmental, provincial or national government or the representatives of state administration. The term also encompasses the representatives of political parties, who are called specifically políticos. Gobierno is further used in a general sense to refer to the representatives of the church or of other non-governmental organizations. In this unspecific general usage people from Huancar conceptualize as gobierno all foreigners connected with a socially recognizable institution which – at least in their opinon – has economic and political regulative power. The Catholic Church is increasingly seen as gobierno because it has developed into an important political actor in north-western Argentina in the last few years. It now manages a considerable portion of the development funds for the highlands. Another aspect of the regulative function of gobierno is that politicians or members of the government are considered to have a certain responsibility for the performance of enterprises. This makes sense if one realizes how closely political and economic capital are interwoven in the Provinces of Jujuy and Salta.

Institutional risks for the pastoral economy in Huancar are in part related to the economic and political instability in north-western Argentina. Among numerous macro-political and macro-economic crises that have arisen during recent years are the introduction of several artificial provincial currencies (three in the last five years), the frequent replacement of governors (eight governors since 1991), with concomitant effects on general policy and composition of public administration, and the rapid succession of insolvencies, and founding of new enterprises in the provincial mining and agro-business sector. Some of the negative impacts of these regional and national crises for households in Huancar are, for example, lack of information about current prices for commodities (especially foodstuffs) and insecurity in respect of labor opportunities in mines and plantations. Another negative consequence emphasized by the district's inhabitants is prevalent uncertainty about reliable interlocutors in public administration and about the functioning of governmental institutions. Ecclesiatical institutions are perceived as unpredictable because of frequent redefinition of their ideological focus and because of unpredictable fluctuations in the intensity of their activities in Huancar. Both aspects are connected with a recent modification of the territorial and administrative structure of the Catholic Church in the highland sector of Jujuy and Salta and with a rather long interregnum in one of the involved episcopates (*Obispado de Humahuaca*).

But not only the absence of a predictable political, capitalistic, and ecclesiastical institutions, which all have a long tradition in the highlands, are perceived as a serious problem by the inhabitants of Huancar. The recent activism of government and church development organizations is also considered to be a source of risks. Because the foreign 'experts' do not take local cultural values and economic practices into account, their meetings and technical proposals are perceived as disruptive. Up to now, no greater development programs have been undertaken in Huancar. The considerable greater presence of *gobierno* is therefore up to now basically a verbal presence. But people already feel disturbed and even embarassed by this verbal presence. The discourses of the different representatives of *gobierno* – politicians, agronomists or catechists – are similar in two points. First, in that they construct an imaginary of life in the highlands (*puna*) by opposing this life to western urban culture. And second, in that they mix up two antagonistic perspectives in this process. On the one hand, they underline the need to 'humanize living conditions for the poor pastoralists, who are abandoned in a harsh environment and are hardly managing to survive'. They argue that in order to achieve this, wild nature needs to be finally 'domesticated'. Technology has to be introduced from more developed areas and new economic practices have to be implemented that maximize production (e.g., enclosures with artificial pastures, weaving machines). The ideas underlying these propositions are that the indigenous inhabitants of this marginal part of the highlands have to be civilized by separating them from nature and that this 'civilizatory process' will make it possible to tame the dangers of nature with technology. The representatives of *gobierno* combine this traditional emphasis on the necessity of a 'civilizing project' for the highlands with a more recent vision of its indigenous inhabitants as the real expert ecologists who live in harmony with nature[6] (some manifestations of this mixture can be seen in the description of a meeting below). The mixture of both lines of argumentation contains numerous incoherences and even contradictions. Thus, in the first line of argumentation 'experts' emphasize the riskiness of nature, while in the second line they talk of the harmony of nature. While on the one hand they point out the necessity of dominating nature with technology, on the other hand they emphasize the advantages of the highlanders' closeness to nature. There is further incongruency in the fact that the 'civilizing' argument is connected with marked paternalism on the part of the 'experts', while the other view postulates alternative forms of agency of indigenous people. This explains why so many people in Huancar complain that through the novel discourses too much pressure is put on them to decide things they do not understand. Up to today they perceive the different development proposals more as a burden than as an opportunity.

Risk-Handling in Huancar: Economic and Social Strategies

A set of structural features of the pastoral economy in Huancar can be interpreted in an abstract way as strategies for coping with environmental and institutional risks (Göbel, 1997 b, 1998 b). They all reflect the interest of the pastoral families in assuring their subsistence basis despite fluctuating conditions. Security aspects play an important role in everyday decision-making. In general, to obtain relatively secure results (e.g., by preferring a familiar and established way of doing things) is given priority over potential short-term production gains. This 'safety-first' logic of the pastoralists sharply contrasts with the emphasis of development 'experts' from government, party and church institutions on the maximization of production through technological investments.[7]

Some of the mechanisms involved in creating security margins are economic diversification and a high degree of economic flexibility. Practices such as mixed composition of flocks, high spatial mobility, a broad spectrum of destinations for trade caravans, or the supplementation of animal husbandry with temporary wage labor or local work arrangements create alternatives to which households can resort in case of necessity. This is not only of great relevance for the handling of environmental risks. It also mitigates the impact of macro-economic crises and prevents falling into excessive dependance on political clientelism.

Another important institution for risk management is the creation and sustaining of social networks (Göbel, 1997 b, 1998 b). They imply moral obligations of mutual help, whose observance can be controlled locally and therefore directly, which is not the case with client-patron relations (e.g., relations of clientelism in the context of political parties). Related to social networks there also exist mechanisms through which the exchange of information can be enhanced, as well as diverse types of intra-household arrangements permitting access to additional labor and resources (pastures, waterholes, animals, animal products, etc.). Thus the ability to fall back upon social networks strengthens the subsistence basis of the household, minimizing not only the impact of ecological events but also the impact of macro-economic and macro-political fluctuations.

Economic diversification and social networks are often emphasized in the literature as central strategies of pastoralists, hunter-gatherers or agriculturalists for coping with environmental risks (Barlett, 1980; Blaikie, Cannon, Davis and Wisner, 1994; Browman, 1987 a, b; Cashdan, 1990; DeGarine and Harrison, 1988; Göbel and Bollig, 1997; Halstead and O'Shea, 1989; Borgerhoff Mulder and Sellen, 1994). However, the handling of institutional risks has so far largely been neglected in anthropological risk analysis. As I have already mentioned, by means of economic diversification and local social networks not only the negative consequences of environmental risks but also those of institutional instabilities can be reduced. Beside this strategy another general mechanism in Huancar for coping with insecurity in asymmetrical relations is the maintenance of an expectant distance towards state administrators, politicians and the church. This attitude is embodied in public arenas by a great spatial distance toward foreigners and by not expressing personal opinions openly (see also the description of the meeting below). Through this peculiar style of interaction herdswomen and temporary labor migrants avoid becoming too dependent (economically and morally) on the state, the church or the administration of mines or plantations. This means they try to keep open the possibility of opting out of 'contracts' at any moment. However, the 'new activism' of *gobierno* in Huancar makes it increasingly difficult for the pastoralists to maintain their traditional distance. In contrast to the more humid northern parts of the highlands of Jujuy, which since the 19th century is much more deeply embedded in the capitalist economy and the state organization, no negotiation strategies have yet been developed in Huancar to cope with the novelty of a considerably greater presence of *gobierno*.

Risk and Cosmology

The heuristic potential of an analytic perspective which takes into account unpredictable fluctuations in the ecological, economic and political conditions, consists in the fact that it enables us to grasp the logic of different economic styles and to see more precisely why and how conflicts arise when different logics clash. The preceding presentation of some features of the economic and social practices in Huancar is necessarily rather synthetic. Nevertheless, it already shows that such an analytical perspective allows to explain why it makes economic sense for the pastoralists to concentrate on assuring their subsistence minimum, to invest in local social cooperative institutions and to maintain the highest possible degree of autonomy. The pastoralists in Huancar do not want to invest too much in one economic, social or political option alone and prefer instead to diversify. In view of the great ecological and institutional risks and the lack of access to monetary capital, greater specialization would considerably increase their vulnerability, in spite of giving them the means of augmenting their production. This is also the reason why they reject the advice of 'experts' from government, political parties or the Catholic Church to concentrate their efforts on one animal species (e.g., through genetic amelioration, enclosures, additional fodder, sanitary measures). As my interviews with the 'experts' show, they use a different economic logic as their starting point. In contrast to pastoralists, 'experts' emphasize not so much security as the maximizing of production. Although 'experts' also lay emphasis on environmental risks, they assume that these risks can be controlled or even overcome through the use of developed technology and greater economic specialization. And some of them mix this argumentation in such a contradictory way with romanticized images of the harmonious union of the highland population with nature, that they are perceived themselves as a risk by the pastoralists.

This general description of differences in the economic styles of pastoralists and 'experts' should suffice to indicate the importance of seeking to understand more precisely the close relationship that seems to exist between risk management and conceptions of nature. For a profounder understanding of the specific features of these economic styles and of the conflicts that arise when they are confronted, environmental perceptions, moral values, and religious beliefs have to be taken into account. As I cannot develop such an analysis here in an extensive form, I will concentrate in the following on one of the major threats to a stock breeder: the death of herd animals. The aim is to show how in the pastoralists' handling of this threat economic and cosmological aspects are deeply interwoven.

During my stay in Huancar in 1996/7 it became very clear to me that this broader perspective, which more strongly contextualizes cultural influences on the perception and handling of risks, is of great relevance to the shaping of development policy. The summer of 1995/96 was extremely dry in the Andean highlands of north-western Argentina. Due to the drought during the whole of 1996 and the beginning of 1997, pumas and foxes appeared unusually often in the vicinity of pastures, homesteads and herding stations. They killed many newborn llamas, lambs, goat kids, sheep and goats in the highlands. Newspapers in the region's political centers of Jujuy and Salta, began to publish sensational stories about the defenseless exposure of humans and animals to these predators in rural areas. Triggered by this media pressure various institutions belonging to the government, the Catholic Church and the political parties announced campaigns against predators in the highlands. In this context they organized also some meetings in Huancar.

The following characterization of such a meeting is based on several detailed observations of clashes between outside 'experts' and the local population of Huancar, which I made from September 1996 until the end of February 1997. All names, that appear in the description, have been changed.

When the representative of the provincial agricultural department, Enrique Sánchez finally reached with his off-road car the village of Huancar in the late afternoon, the village assembly had almost broken up again. People had waited patiently all day for this man, who had announced his visit by means of an official letter from the governing party. The agricultural expert hurried into the meeting room, where the men were sitting on one side and the women and small children on the other. Using the quick and direct language of the city, the man spoke of the necessity, in view of the catastrophic drought and high animal losses, for organizing communal hunting expeditions against pumas and foxes. He proposed that the president of the village assembly should request the party to provide rifles, ammunition and traps. Then he started to say, that he really wants to help the people of Huancar wherever possible, since the members of his party believed that the only way to protect the herds of the poor highlanders was by taking drastic action against the predators. They saw it as their duty to help the people here (*la gente de acá*) to a more humane way of life, so that they might become more developed (*progresar*). For they were the true owners of this land (*dueños de la tierra*) and they knew how to survive in such a dry region, a desert (*saben como vivir en una parte tan seca de la puna*). They were the true ecologists (*son los verdaderos ecologistas*). Don Eustaquio Quispe, the elected president of the village assembly, stood in the middle of the room and thanked Enrique Sánchez at length for his sympathy and his concern. The other men and women remained silent and stared blankly in front of them. In the late evening, long after we had returned to the farm, we were talking over the events of the day. Suddenly Doña Eusebia said: 'How could we just go out and kill the fox or the puma?' That would put our animals in even greater danger. Through such an action luck can be damaged (*se puede dañar la suerte*).

Doña Eusebia's reaction to the 'expert's' proposal may seem puzzling at first glance. But if we take into account the relationship between risk handling and environmental perception in Huancar, it is culturally a very coherent reaction. In order to understand why the killing of a fox or a puma can have such negative consequences for the whole herd and is in general not an adequate strategy, it is necessary to present some core elements of the pastoralists' cosmology and their relationship to the environment.

The Socialization of Nature

The pastoral *Lebenswelt* is characterized by a close relationship between humans and the environment. In indigenous conceptualizations of nature, patterns of human social life and pastoral production play an important role. Nature, or at least the majority of its components, is not perceived as being completely outside of the human sphere. Herding stations, for example, which usually consist only of a windbreak and the surrounding pasture grounds and waterholes are considered to be female spaces; as spaces which, to put it more more abstractly, are socialized by herdswomen and thus belong firmly to the human sphere. Closely linked to this is the emotional relationship between a herdswoman and her herd animals which is expected of her by societal norms. This relationship is characterized by certain kinds of prescribed social behavior which are reminiscent of the way a mother is expected to behave towards her children. Frequently heard comments are: 'You have to look after a herd animal, think of it, like it, not get irritable and loud with it, talk to it calmly, keep it under control but allow it some freedom'. This behavior is appropriate since llamas, sheep and goats also possess a soul (*alma*), feelings (*sentimiento*) and intelligence (*mentalidad*). Thus, as with humans, it is important to respect their individuality, which is expressed through having a name, special characteristics and their own life story. The same applies to dogs and cats, which, together with llamas, sheep and goats, are the domestic animals with which humans have the closest contacts in their daily life. These five animals are therefore referred to as true animals (*animales*). The way human characteristics are attributed to these animals, however, suggests that calling them animals is not intended to emphasize their separation from the human sphere as much as to indicate their social status within the household community.

People in Huancar relate to nature through *pachamama*, or Mother Earth. There is great variation in the way Mother Earth is conceptualized and in the aspects which are emphasized according to the concrete situation. On the one hand *pachamama* is conceived of as an abstract entity that maintains the life cycles, creating and destroying life. On the other hand, however, she is also conceived of as a concrete entity that is localizable in space and has human characteristics. She is concrete because people stand on her. In this sense she is also called *tierra* (earth). She can also be concretely localized in namable places, such as waterholes, sand dunes, volcanoes and certain hollows in the ground (*tierra pesada*). All these are openings of the *pachamama* in which herd animals and humans can be swallowed up. Such openings are therefore considered to be dangerous places. It is not thought of as a contradiction that beside this spatial and material concept of *pachamama*, she is also described as a family, consisting of father, mother and child (*pachatata, pachamama* and *pachaguagua*). She is always spoken of in the singular, and a large number of metaphors are used which refer to human characteristics such as '*pachamama* can be angry' or 'happy', 'sad', 'hungry' or 'thirsty'.

Like humans, *pachamama* also keeps domestic animals: the fox is her dog, the puma (*Felis concolor*) her cat, vikuñas (*Lama vicugna*) are her llamas, chinchillas (*Chinchilla brevicoudata*) her sheep and viscachas (*Lagidium viscacha*) her donkeys. She has no goats, since these come from the devil. As humans do, also *pachamama* herds her llamas and sheep, castrates them and puts marks on their ears as a sign of ownership. Like the herds of humans, her herd animals live in localizable pasture grounds with enclosures and houses, only that these are not in the valleys, where the humans live with their animals, but are usually on the hilltops.

The closely woven fabric of interaction between humans and their environment has both a harmonious and a conflicting side. One central element guiding the relationships with nature is the element of reciprocity. People in Huancar think, that resources should only be exploited or herd animals only be killed if at the same time something is given back (e.g., in the form of offerings to the pachamama or an intensive emotional engagement with other herd animals). In daily economic practices they also care very much not to use more resources than they inmediately need. An excessive usage would mean to disrupt harmony with nature.

At the same time the pastoralists stress manigfold dangers that can emanate from nature. Disharmony with nature arises mainly from the ambivalent actions of *pachamama*, Mother Earth. According to the understanding of the highlanders, economic success is due only partially to their own efforts. They refer to economic success in pastoral production with the term *suerte* (luck). The most decisive factor for having or not *suerte* is the goodwill of *pachamama*. While on the one hand she can exercise a positive influence on the growth of the herd, on the other hand she can also swallow animals up or has the power to reduce severely their productivity. In order to invoke the goodwill of *pachamama*, humans carry out special sacrificial rituals in the course of which they present her with gifts (e.g., coca leaves, alcohol, food). As will be described below in greater detail, they also try to observe the moral prescriptions for 'correct' behavior in respect of the environment, in order not to incur her anger.

Why Herd Animals Die

Western-minded 'experts' from state, church and party organizations tend to see only physiological factors (illness or external violence) as being responsible for the death of a herd animal. They believe that these physiological factors are related to the poor conditions in which the animals are kept (e.g., deficient sanitary conditions, underfeeding, poor genetic base of stocks), to extreme climatic events (e.g., snow), or to the activities of predators. The pastoralists, on the other hand, take a much broader range of causal factors into account. They also base their views on quite different causal link patterns – see Figure 9.2.

Just like the 'experts' the pastoralists believe, that an animal may be killed directly by climatic events such as lightning (*rayo*), hail (*granizo*), snow (*nieve*) or drought (*sequía*), as well as by poor pasture conditions (e.g., by harmful plants – *malpasto*), or by a puma or a fox (*zorro*). They also mention, like the 'experts', a number of indirect causes. Thus an animal might become ill (*se enferma*) and might die because it has caught cold in a hail or snow storm, or because it became weak when the pastureland shrivelled up during a drought or he has eaten harmful plants. However, the highlanders relate these factors to different cause and effect chains and understand them as a complex, coherent whole. This means that unlike the 'experts' they see no fundamental break between climatic events such as a hail shower and the appearance of social envy (*envido*) in the

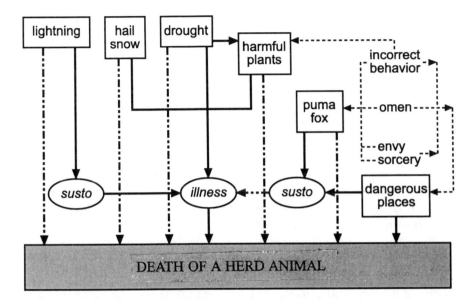

Figure 9.2   Causalities for the death of a herd animal

form of a puma, for example. Moreover, the causal factors they mention have a different cultural base. Thus, they emphasize that illness can be triggered not only by climatic events, insufficient or bad food, but also and especially by *susto* (fright). If an animal or a person is suddenly frightened, then the soul leaves the body. As a consequence, certain organs are affected. An animal can be frightened if lightning strikes close to it, if it passes near a 'dangerous place' (*lugar peligroso*) such as a waterhole or certain hollows in the ground, or if it sees a fox or a puma. But what makes waterholes or certain shapes in the ground 'dangerous places' or why do pumas or foxes cause *susto*? To answer these questions we must look at other possible causes which take us even further into the field of cosmology: incorrect behavior, omens, envy, magic. To demonstrate the significance of these elements, I will use the example of the different ways in which a puma or fox may endanger the herd animals .

Depending on the manner in which a puma or a fox approaches a herd and how it behaves towards the animals, it is considered either as a wild animal, a domestic animal belonging to *pachamama*, an omen, a medium for social envy or as a medium for an act of sorcery. The distinction is important because it determines the way in which the pastoralists react to the loss of a herd animal. Thus one characteristic of environmental cognition in Huancar is flexibility in classifying natural phenomena. This flexibility leads not only to different denominations for the 'components of nature' (e.g., plants, animals) but also implies different modes of interaction with the components named.

Pumas and foxes are considered to be wild animals (*animales salvajes*) which can damage the herd directly (*hacen daño*) if they creep up to a single llama, sheep or goat which is grazing apart from the herd or which has got lost. This 'wild' behaviour on the part of the puma or fox is due to hunger, and humans may react by hunting it down. Only in this context is it possible for the pastoralists to consider seriously the proposal of the 'experts' to hunt predators with weapons and traps, and then only if the hunt is limited to that puma or fox which is believed to be responsible for the loss of a herd animal. It is only in this context that hunting the puma or fox is not likely to bring about further danger for the herd.

However, if during the night a puma or fox enters the enclosure of the sheep and goats or goes to the place where the llamas sleep at night, which means coming close to humans and their fireplaces, and if it kills a large number of the best herd animals but leaves most of them lying on the ground, then it is not behaving like a 'wild' animal, but as a 'domestic' animal, in the sense of a domestic animal belonging to *pachamama* (Mother Earth). *Pachamama* has sent the puma or fox because the members of this household are guilty of incorrect behavior (*mal comportamiento*): They may have killed too many of *pachamama*'s domestic animals, meaning they have hunted too many vicunas or viscachas without asking her permission first or without being driven by hunger; they may have failed to make enough sacrificial offerings to her, or maybe they have neglected their llamas, sheep and goats.

Another interpretation for the killing of many herd animals is that the puma or fox is a medium for social envy (*envido*) or for a charm (*hechizo*). In such a case, the predator will carry off most of the animals. A person from within the social network of the members of the household, usually a close relative, is envious of the growth of the household herd and has therefore mentally dispatched an *envido*, or has employed sorcery (*hechizo*) through a ritual, in order to attack the intended target in the form of a puma or fox.

If a fox or puma approaches the herd during the day, in full view of everyone, and sits for hours watching the animals from a short distance or plays with them by running around them, then it is an omen (*aguero*). It is the carrier of an *ánimo*, the spirit-soul of a friend, *compadre* or *comadre* (ritual kins), close relative of a member of the household, who wishes to take leave of him before his approaching death. An *ánimo* may also tease the person, for example by killing a few of his herd animals. But because it is an *ánimo*, there is no possibility of taking action against the puma.

Cultural Risk Management: The Reproduction and the Restitution of Order

How can the death of a herd animal be prevented? What can be done in general to guard against *peligros* and *problemas*? What can be done to prevent that a damage does not go worse?

The pastoralists distinguish between direct action to cope with the concrete impact of a danger, and general preventive measures against risks which serve to ensure *suerte*, or economic success. By respecting certain behavioral norms, each individual hopes to secure for himself the benevolence of *pachamama* and prevent in this way economic setbacks. The correct management of the environment is here of central importance. One of the rules of proper environmental management is moderate use of resources, based entirely on the premise of necessity. Killing a wild or domestic animal or plucking a plant just for pleasure is considered to be wrong, and therefore dangerous, behavior. People should use natural resources only when they are pressed by the necessity (*necesidad*) of satisfyfing their basic needs, such as hunger. A second rule concerns bans which have to be observed. Some plants and animals may not be used for economic purposes at certain times. For example, herd animals should not be slaughtered on Tuesdays or Fridays, since these are dangerous days and doing so might have a negative impact on *suerte*. The utilization of some plants and animals is even completely taboo. For example, gathering cactus flowers is forbidden and the Andean ostrich *suri* (*Pterocnemia pennata*) must not be hunted. A third rule concerns the proper control of herd animals. Herders must not neglect their herd animals; they should look after them, like them, and not get irritable or loud with them. It is also important to treat the means of successful breeding with care, as a preventive measure against losses. The fertility of a herd, in the sense of its growth potential, is concentrated in the breeding animals, i.e., the adult llama stallion, ram and billy goat, as well as in certain body parts of all herd animals (e.g., in parts of the ribs and in the fat). Breeding animals and these body parts must therefore be treated with special care. They may not be given to anyone else, for that would be equivalent to giving away one's breeding success.

In addition to correct treatment of the environment, sacrificial offerings to *pachamama* are also of great significance as a general preventive measure against risks. It is important that these ritual ceremonies are carried out with conviction and that the offerings of coca, alcohol and tobacco are seen to be generous. Guests who are invited to the ceremony must also be well plied with food and drink.

The careful observation of omens is a further preventive measure. It is important to be attentive and receptive to the various portents which can signalize approaching danger (dreams, the changing winds, birds, insects). In this way the herder prevents to be exposed – at least in psychological terms completely unprepared – to a loss. Thus, the negative impact of a threat is kept within limits. As he was forewarned, the herder does not get such a shock when he finds one of his animals dead, which means that he himself is then in less danger of becoming ill through *susto*.

The rules for correct behavior outlined above are not directed against any particular danger but serve to reduce the likelihood of losses occurring. At the same time, it is also possible to take direct action against certain risks, in the sense of minimizing the damages once they have occured. The aim here is to reestablish the equilibrium between humans and Mother Earth, between *suerte* and herder, after it has been upset. The worsening of the situation is therefore prevented. For the sake of simplicity I shall again use the example of the puma to demonstrate this. If incorrect behavior was the reason why the puma has killed a herd animal then the herder must beg *pachamama* to forgive him for his wrong behavior, for example for killing one of her domestic animals, and he must pay for this with generous sacrificial offerings (*ch'allas*) of alcohol, coca leaves, animal fetuses, animal heads or meat. But he should not hunt the puma. However, if the reason was *susto* or another illness, then a healer (*curandero*) must be hired to carry out a healing ritual for the sick animals. But if the puma was sent out by means of sorcery due to social envy, then the danger must be neutralized by counter-sorcery (*vuelto*), carried out by a specialist.

Conclusions

The preceding discussion of how pastoralists in Huancar handle one specific risk – the death of an herd animal – already hints at same basic differences that exist between them and the 'experts' concerning risk perception in general. These differences in risk perception are an expression of the distinct *Lebenswelten* in which pastoralists and 'experts' are embedded. The worldviews connected with the distinct domains of experience of the pastoralists and the 'experts' structure their relationships towards animals and the natural environment in specific ways. So we have seen that in the indigenous worldview the human and the natural spheres are closely interwoven. The **pastoralists'** close relationships with the environment and in particular with their herd animals can be described as very personal or even intimate, although this does not mean that they do not take the concrete utility of an animal as a resource into account. Relationships to nature are mediated by *pachamama*. Each individuum tries to build up in the course of its life strong webs of reciprocity towards *pachamama* in order to assure economic success. The attitudes of the representatives from state, party or ecclesiastical institutions towards nature are characterized by a much greater distance. Contrary to the pastoralists, they perceive specific resources (e.g., pastures, waterholes, domestic animals) as isolated items to which they refer mainly in economic terms. Human interactions with these resources, as for example animal husbandry, are conceptualized to a great extent only in productive terms (e.g., quantities of meat and wool, monetary values). The 'expert's' relationship to nature is basically mediated by technology, that has not been developed locally but in different ecological, economic and cultural contexts. The 'experts' stress mainly ecological risks for pastoral production in the highlands. They see these risks as central threats to the productivity of animals; focusing their analysis on the future economic consequences of ecological risks. Therefore they propose, that these risks should be controlled by means of technical investments and a greater productive specialization, both aiming at a maximization of animal productivity.The pastoralists, on the other hand, emphasize that major threats to their economy are not only the dangers, which can emanate from the environment. Besides ecological risks they underline also the inpredictabilities of *gobierno*. In regard to their handling of ecological risks, we have seen, that they not only relate the ocurrence of a risk but also the possibilities to prevent a risk or to delimit its impacts to their previous performances with nature and society. In the pastoral worldview risk management has to take past interactions with *pachamama*, plants and animals, the own domestic animals and with other families into account. The underlying idea is, that many threats of the present are caused by incorrect, and therefore dangerous behaviors of the past. Thus, one specificity of the herder's conceptualization of risks is that they define them in other time scales than the 'specialists'. The other characteristic is that in their perspective the handling of risks has not only a concrete economic but also a religious and social dimension.This makes clear why development measures that do not take these differences in environmental perception and risk management into account are condemned to failure.

Notes

[1]
I carried out two extensive field studies in the District of Huancar from September 1991 to August 1993, and from September 1996 until the end of February 1997. On different occasions I have had the opportunity to return to the area for shorter stays (March-April 1994, July-August 1994, July 1995, June-July 1997). I further undertook a comparative field study from July to September 1998 in different parts of the Andean highlands of north-western Argentina and in the city of Jujuy.

[2]
Households are referred to as *familias*. People differentiate between families that live from livestock-breeding (*'familias que pastorean hacienda'*) and families that do not raise animals (*'familias que ya no viven de la hacienda'*). Two aspects are central to the cultural definition of pastoral households. One variable is the right to share pastoral infrastructure and activities. Those people classed as members of a pastoral household use the same pastures and waterholes, herd animals jointly, cook together at one fireplace, eat from one cooking-pot and sleep in the same herding camps or at least are allowed to do so due to their bilateral filiation and their affinal ties. The other variable for the determination of a pastoral household is the identifiability of its members with the locus of a homestead.

[3]
In 1993, the heads of six of these non-pastoral households were public service workers (school, road construction), received a regular pension, and/or owned small trade stores. Their economic reorientation is a relatively new development and dates from between 1986 and 1991. The remaining two non-pastoral households (in 1993) did not have such stable income alternatives, and were therefore very poor. They lived on income from the production of knitted goods (e.g., socks, gloves) and irregular government subsidies. These two households could no longer herd animals, because family members were old and sick. In 1996, all members of one of these two households had died. Another difference as compared to the situation in 1993 was that one pastoral household had migrated, while two new households of public service workers had been created. In this article, I will concentrate mainly on the pastoral households in Huancar.

[4]
Referring to other authors, Yates and Stone (1992, p. 4) state that '... risk is the possibility of loss, ... the existence of threats'. For them '[t]he critical elements of the risk construct are (a) potential losses, (b) the significance of those losses, and (c) the uncertainty of those losses'. And Renn (1998, p. 50-1) writes in a recent review of risk research from technical, economic, psychological and sociological standpoints: 'All risk concepts have one element in common ...: the distinction between reality and possibility. If the future were either predetermined or independent of present human activities, the term "risk" would make no sense. ... [It] is often associated with the possibility that an undesirable state of reality (adverse effects) may occur as a result of natural events or human activities. ... Risk ... includes the analysis of cause-effect relationships ..., but it also carries the implicit message to reduce undesirable effects through appropriate modification of the causes or, though less desirable, mitigation of the consequences.'

[5]
The cultural aspects of risk perception in Huancar were mainly studied by means of participant observation and interviews. In addition, people were given cards with the risk categories, that I elucidated from observations and interviews, in order to focus more directly on risk evaluation and causal explanations of risks. For ranking methods

applied (e.g., pile sorting) see Bernard, H.R. (1994), *Research Methods in Cultural Anthropology: Qualitative and Quantitative Approaches,* Sage, London.

6   The image of 'the ecologically noble savage' (Redford, 1992) is the latest expression belonging to a long-standing tradition in western thought of idealizing the 'primitive Other' (Conklin and Graham, 1995; and compare the German concept of *Naturvolk*). The global diffusion and popularization of this image was triggered by Agenda 21 (United Nations Rio Conference, 1992) and reinforced by posterior developments in (western) environmentalistic ideology. Not only the discussion regarding sustainable development and biodiversity, but also the debate on the subject of indigenous intellectual property rights rely heavily on the vision of indigenous people as natural conservationists (for a bibliographic overview see, for example, Antweiler, 1998; Conklin and Graham, 1995; Orlove and Brush, 1996; Milton, 1993, 1996; Sponsel, 1995). In recent years these ideological developments also reached the political centers of north-western Argentina. The Argentine constitution was reformed in 1995. For the first time ample concessions were made to indigenous communities: land-titles, exclusive rights to utilize all the resources, and protection of indigenous knowledge systems. Although up to now (October 1998) no legal instruments exist in the Province of Jujuy to implement these rights, the reform created political pressure and favorable public opinion for undertaking actions with a certain 'ethno-environmentalistic touch' and for devoting development funds to such actions.

7   The predominance of the 'safety-first principle' in peasant economies and its relevance for the comprehension of colonial and post-colonial transformation processes has been underlined in moral economy (e.g., Scott, 1976; Watts, 1988) and in risk analysis from the point of view of decision theory (e.g., Barlett, 1980; Cashdan, 1990; Halstead and O´Shea, 1989; Ortiz, 1980). A synthesis between moral economy and decision theory was suggested by Godoy (1985). For recent analyses of the clashes of different economic logics in developmental matters, see Hess, C. (1997), *Hungry for Hope: On the Cultural and Communicative Dimensions of Development in Highland Ecuador,* Intermediate Technology Publications, London.

References

Antweiler, C. (1998), 'Local Knowledge and Local Knowing: An Anthropological Analysis of Contested "Cultural Products" in the Context of Development', *Anthropos,* vol. 93, pp. 469-94.

Barlett, P. (ed.) (1980), *Agricultural Decision Making: Anthropological Contributions to Rural Development,* Academic Press, New York.

Bianchi, A. (1981), *Precipitaciones en el Noroeste Argentino,* Instituto Nacional de Tecnología Agraria, Salta.

Blaikie, P.T., Cannon, I., Davis, I. and Wisner, B. (1994), *At Risk: Natural Hazards, People's Vulnerability and Disasters,* Routledge, London.

Bollig, M. (1997), 'Risk and Risk Minimization among Himba Pastoralists in Northwestern Namibia', in B. Göbel and M. Bollig (eds), *Risk and Uncertainty in Pastoral Societies,* Nomadic Peoples (NS) 1, Berghahn, Oxford, pp. 66-89.

Borgerhoff Mulder, M. and Sellen, D.W. (1994), 'Pastoralist Decisionmaking: A Behavioral Ecological Perspective', in E. Fratkin, K. Galvin and E.A. Roth (eds), *African Pastoral Systems: An Integrated Approach,* Lynne Rienner, London, pp. 205-29.

228 Coping with Changing Environments

Browman, D. (1974), 'Pastoral Nomadism in the Andes', *Current Anthropology*, vol. 15, pp. 188-96.

Browman, D. (1987 a), 'Agro–pastoral Risk Management in the Central Andes', *Research in Economic Anthropology*, vol. 8, pp. 171–200.

Browman, D. (ed) (1987 b), *Arid Land Use Strategies and Risk Management in the Andes*, Westview Press, Boulder.

Brush, S. and Guillet, D. (1985), 'Small-scale Agro-pastoral Production in the Central Andes', *Mountain Research and Development*, vol. 5, pp. 19-30.

Cashdan, E. (ed) (1990), *Risk and Uncertainty in Tribal and Peasant Economies*, Westview Press, Boulder.

Conklin, B. and Graham, L. (1995), 'The Shifting Middle Ground: Amazonian Indians and Eco-politics', *American Anthropologist*, vol. 97, pp. 695-710.

DeGarine I. and Harrison, G.A. (eds) (1988), *Coping with Uncertainty in Food Supply*, Clarendon Press, Oxford.

Delgado, F. and Göbel, B. (1995), 'La historia olvidada de la Puna de Atacama', in M. Lagos (ed), *Jujuy en la historia, Avances de investigación II*, Universidad Nacional de Jujuy, Jujuy, pp. 117–42.

Escobar, A. (1995), *Encountering Development: The Making and Unmaking of the Third World*, Princeton University Press, Princeton.

Fischhoff, B., Lichtenstein, S., Slovic, P., Derby, S. and Keeney, R.L. (1981), *Acceptable Risk*, Cambridge University Press, New York.

Flores Ochoa, J.A. (1968), *Los pastores de Paratía*, Instituto Indigenista Interamericano, Lima.

Fratkin, E. (1997), 'Pastoralism: Governance and Development Issues', *Annual Review of Anthropology*, vol. 26, pp. 235-61.

Fukui, K. and Markakis, J. (eds) (1994), *Ethnicity and Conflict in the Horn of Africa*, James Currey, London.

Göbel, B. (1997 a), 'Geschlecht und Produktion: Hirtinnen im Andenhochland Nordwest-argentiniens', in G. Völger (ed), *Sie und Er, Frauenmacht und Männerherrschaft im Kulturvergleich, Materialband zur Sonderausstellung in der Josef-Haubrich-Kunsthalle*, Köln, pp. 367-74.

Göbel, B. (1997 b), '"You have to Exploit Luck": Pastoral Household Economy and the Cultural Handling of Risk and Uncertainty in the Andean Highlands', in B. Göbel and M. Bollig (eds), *Risk and Uncertainty in Pastoral Societies*, Nomadic Peoples NS (1), Berghahn, Oxford, pp. 38-53.

Göbel, B. (1998 a), 'Räume geschlechtlicher Differenz und ökonomisches Handeln (Andenhochland Nordwest-Argentiniens)', in B. Hauser-Schäublin and B. Röttger-Rössler (eds), *Differenz und Geschlecht: Neue Ansätze in der ethnologischen Forschungsperspektive*, Reimer Verlag, Berlin, pp. 136-62.

Göbel, B. (1998 b), 'Risk, Uncertainty and Economic Exchange in a Pastoral Community of the Andean Highlands (NW-Argentine)', in T. Schweizer and D. White (eds), *Kinship, Networks and Exchange*, Cambridge University Press, Cambridge, pp. 158-77.

Göbel, B. (1998 c), '"Salir de viaje": Producción pastoril e intercambio económico en el noroeste argentino', in Dedenbach-Salazar Sáenz, S. Arellano Hoffmann, C., König, E., Prümers, H.(ed), *50 años de estudios americanistas en la Universidad de Bonn: Nuevas contribuciones a la arqueología, etnohistoria, etnolingüística y etnografía de las Américas, 50 Years Americanist Studies at the University of Bonn: New Contributions to the Archaeology, Ethnohistory, Ethnolinguistics and Ethnography of the Americas*, Bonner Amerikanistische Studien No. 30, Anton Saurwein, Markt Schwaben, pp. 867-91.

Göbel, B. and Bollig, M. (eds) (1997), *Risk and Uncertainty in Pastoral Societies*. Nomadic Peoples (NS) 1, Berghahn, Oxford.

Godoy, R. (1985), 'Risk and Moral Contract in Bolivian Peasant Mining', *Research in Economic Anthropology*, vol. 7, pp. 203–24.

Halstead, P. and O'Shea, J. (eds) (1989), *Bad Year Economics: Cultural Responses to Risk and Uncertainty*. Cambridge University Press, Cambridge.

Kasperson, R.E. and Kasperson, J.X. (1987), *Nuclear Risk Analysis in Comparative Perspective*, Allen and Unwin, Winchester.

Milton, K. (ed) (1993), *Environmentalism: The View from Anthropology*, Routledge, London.

Milton, K. (1996), *Environmentalism and Cultural Theory: Exploring the Role of Anthropology in Environmental Discourse*, Routledge, London.

Orlove, B. (1981), 'Native Andean Pastoralists: Traditional Adaptations and Recent Changes', in P. Salzman, *Contemporary Nomadic and Pastoral Peoples: Africa and Latin America*, Studies in Third World Societies No. 17, Department of Anthropology, Williamsburg, pp. 95-136.

Orlove, B. and Brush, S. (1996), 'Anthropology and the Conservation of Biodiversity', *Annual Review of Anthropology*, vol. 25, pp. 329-52.

Ortiz, S. (1980), 'Forecast, Decision, and the Farmer's Response to Uncertain Environments', in P. Barlett (ed) *Agricultural Decision Making: Anthropological Contributions to Rural Development*, Academic Press, New York, pp. 177-202.

Palacios Ríos, F. (1981), 'Tecnología del pastoreo', in H. Lechtman and A.M. Soldi (eds), *La tecnología en el mundo andino*, Universidad Nacional de México, México, pp. 217-32.

Quinteros, H.O. and Masuelli, R. (1991), *Censo '91, Para darnos cuenta*, Dirección Provincial de Estadísticas y Censo, Jujuy.

Redford, K. H. (1990), 'The Ecologically Noble Savage', *Orion Nature Quarterly*, vol. 9, pp. 25-9.

Renn, O. (1998), 'Three Decades of Risk Research: Accomplishments and New Challenges', *Journal of Risk Research*, vol. 1, pp. 49-71.

Scott, J. (1979), *The Moral Economy of the Peasant: Rebellion and Subsistence in Southeast Asia*, Yale University Press, New Haven.

Sponsel, L. (1995) 'Relationships among the World System, Indigenous Peoples, and Ecological Anthropology in the Endangered Amazon', in L. Sponsel (ed), *Indigenous People and the Future of Amazonia: An Ecological Anthropology of an Endangered World*, University of Arizona Press, Tucson, pp. 263-94.

Troll, C. (1968), 'The Cordilleras of the Tropical Americas: Aspects of Climatic, Phytogeographical and Agrarian Ecology', in C. Troll (ed), *Geo-ecology of the Mountainous Regions of the Tropical Americas*, Dümmlers, Bonn, pp. 15-56.

Watts, M. (1988), 'Coping with the Market: Uncertainty and Food Security Among Hausa Peasants', in I. DeGarine and G. Harrison (eds), *Coping with Uncertainty in Food Supply*, Oxford University Press, Oxford, pp. 260-89.

West, T. (1983), 'Family Herds – Individual Owners: Livestock Ritual and Inherit-ance among the Aymara of Bolivia', in R. Berleant-Schiller and E. Shanklin (eds), *The Keeping of Animals: Adaptation and Social Relations in Livestock Producing Communities*, Allanheld and Osmun, Totowa, pp. 93-106.

Yates, F. and Stone, E. (1992), 'The Risk Construct', in F. Yates (ed), *Risk-Taking Behavior*, John Wiley, Toronto, pp.1-25.

10 Risks and Coping Strategies of a Vulnerable Group in the Dominican Dry Forest
The case of charcoal burners in Chalona

VERONIKA ULBERT

Introduction

The widespread destruction of tropical forests on the Caribbean Islands and resulting environmental problems such as soil erosion, decrease of soil fertility and the reduction of varieties in flora and fauna has increasingly become a matter of scientific investigation in the Dominican Republic since the beginning of the 1990s (García, 1994; Basilis, 1995; Bolay, 1997; May, 1997). What, however, has been comparatively less regarded are the social consequences emerging from advancing deforestation.

Hence, the following contribution shall investigate what the consequences of deforestation of tropical dry forests areas in the south-west of the Dominican Republic are for such human groups whose livelihood system decisively depends upon these forest resources. The question is whether these people, who already suffer from difficult socio-economic conditions, are able to cope with the deterioration of their environment by applying risk reducing strategies, or whether the destruction of natural resources will be a reason for further social vulnerability of the wood worker families.

According to Chambers (1989), 'vulnerability' is the opposite term of 'security'. It includes dimensions that go beyond the question of economic poverty of social groups defined as shortage or need. Consequently, the dimensions encompass their exposure to risk, their social defencelessness and dependence, their political powerlessness as well as their humiliation and psychological harm.

To comprehend the degree of vulnerability of Dominican dry forest inhabitants affected, the article will focus on the following questions.

231

'Exposure': What are the existential risks of this human group and what can be defined as reasons of exposure?

'Capacity': What strategies are available to Dominican dry forest inhabitants in order to cope with processes of external destabilisation? To which extend are they supported by development programmes?

'Potentiality': What is the potential of the livelihood system of dry forest inhabitants to recover after environmental destruction? By a systematic analysis of social vulnerability as done here, the methodological aspects of human ecology, expanded entitlements and political economy will overlap (Bohle, Downig and Watts, 1994).

Under the terms specified, it will be of particular interest to investigate the quality of dry forest resources considering both their value for human groups and their natural recovery potential. The focus will also be on the access conditions to resources which are existential for the survival of dry forest households. It will also take into consideration 'net assets' (Chambers, 1989) which dry forest inhabitants themselves contribute to the improvement of their survival chances. Moreover, the social and legal status of dry forest inhabitants within the total population of the Dominican Republic will be examplified. Finally, this article will point out to what degree socio-psychological factors will influence the vulnerability of human groups within endangered ecosystems. It was noted by Chambers (1989, p. 1) that precisely these factors tend to be easily underestimated if the scientific analysis does not provide the affected marginal group with a chance sufficient enough to define by themselves along which lines they consider them to be be vulnerable. He states that

> care is also needed because vulnerability and security start as 'our' concepts and are not necessarily 'theirs'. To correct and modify them to fit local conditions requires decentralised analysis, encouraging, permitting, and acting on local concepts and priorities, as defined by poor people themselves.

Forest Destruction in the Dominican Republic

Forest destruction is one of the most pressing ecological problems of rural Dominican Republic. The areas most affected by deforestation are mountainous regions which represent about 60% of the Dominican national territory.

Whereas in mid 20th century, an estimated 50% of the country was covered by forests, it had decreased to 10% of the country's total surface in 1996 – see Figure 10.1. There is reported evidence that over the past 70 years an estimated 430 km² of Dominican forests have been cut down annually. This is extremely alarming on the background that from soil-specific criteria it is suggested that approximately 56% of the nation's total area should be reserved specifically for forestry purposes (Perelló-Aracena, 1991; Informe Nacional, 1991).

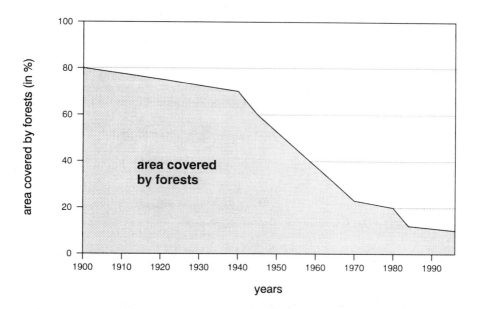

Figure 10.1 Extent of deforestation in the Dominican Republic

Source: Ceara (1987), *Land Tenure and Agroforestry*, Nairobi; Perelló-Aracena (1991), *Energy*, Idstein; Comisión Nacional Técnica Forestal (1991), *Plan de Acción Forestal para la República Dominicana*, CONATEF, Santo Domingo; Centro de Investigación y Promoción Social (1996), *República Dominicana: Un País en Procesos de Cambio? Coyuntura Económica, Social, Política y Cultural*, CIPROS, Santo Domingo; Munzinger Archiv (1996), *Dominikanische Republik*, Ravensburg.

The systematic exploitation of Dominican forest resources can be traced back to the first half of the 19th century when national and international wholesalers started to export tropical precious wood primarily to Europe (Cassá, 1991; Bolay, 1997). At present, Dominican forests are primarily exploited in their function as energy resources due to the increasing national demand for fuelwood and charcoal. Also, each year extensive areas of forest land disappear in mountainous areas due to

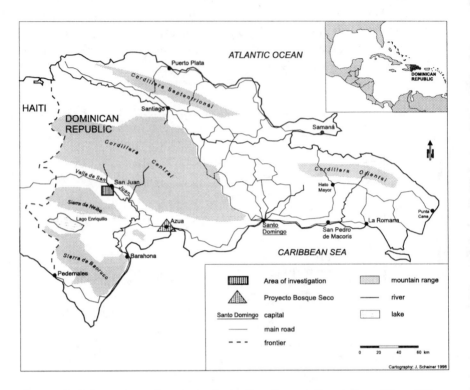

Figure 10.2 Location of case study in Southwest Dominican Republic

clearing activities of smallholder farmers in an attempt to gain new arable land. This occurs on the one hand as a result of the decrease of soil fertility due to many years of agricultural cultivation. On the other hand, deforestation could be seen as a result of the lack of fertile arable land in low-situated regions as a consequence of property concentration there (May, 1997).

Current deforestation in the Dominican Republic has resulted in a steady decline of the country's natural biodiversity. It is estimated that almost 300 native species of Dominican flora and fauna are threatened with extinction as a direct consequence of forest destruction (Ceara, 1987). Especially in mountainous regions with short, intensive rainfalls and deforestation causing severe soil erosion, not only migratory agricultural systems but also water-catchment areas being important to national water supply are affected. As a matter of fact, nearly 86% of those soils are considered to be seriously degraded due to soil erosion (Informe Nacional, 1991).

Dry Forests in the Dominican South-West

The south-western part of the Dominican Republic covers seven provinces which together make up for about 30% of the national territory. The area as given in Figure 10.2 stretches from the Cordillera Central in the north to the Caribbean Sea in the south and to the Haitian border in the west.

In comparison to other parts of the country, the Dominican south-west is considered to be the area most affected by deforestation. Between 1964 and 1990, about one third of the region's forests disappeared (Deutsche Gesellschaft für Technische Zusammenarbeit, 1994). Extensive mahogany populations as well as other valuable species in the area became victims of last century's timber-exports. Since that time, the forests of the south-west have never regenerated to their original state, i.e., a forest with high degrees of different species (Bolay, 1997; Oehlschläger, 1996). Considering that tropical precious wood is endangered in this area, less valuable species used in the production of timber, fuelwood and charcoal are harvested from tropical dry forests, which cover about 35% (or the equivalent of 550,000 ha) of the south-western region (Oehlschläger, 1996).

The south-western dry forest (*Bosque Seco*) is made up of secondary vegetation as a result of earlier human intervention. It extends from the coast up to an altitude of about 400 m and is located in the rain shadow of high mountain ranges (Coordillera Central, Sierra de Neiba, Sierra de Baoruco) which function as natural obstacles against the rainy north-eastern trade winds. Depending on altitude and geographical position the amount of annual rainfall varies from 350 to 700 mm distributed over two rainy seasons. Between 6 to 8 months of the year, the climate is extremely dry. Medium annual temperatures range from 25 to 28°C (Erhart, 1995).

The semiarid climatic conditions require special adaptation of dry forest ecosystems. The majority of trees are characterised by leaves which are small, relatively delicate and adapted for storing water. The leaves are sometimes shed during the dry season. In comparison with evergreen tropical rain forests, dry forests are characterised by a relative lack of species variety. The representative species of trees, among others, are Bayahonda (*Prosopis juliflora*), Guyacan (*Guajacum officinale*), Baitoa (*Phyllostylon brasiliensis*), Cambrón (*Acacia macrantha*) and different species of cactus, such as Caguey (*Neoabottia paniculata*) or Alpargata (*Opuntia moneliformis*). Considering local hydrographic conditions and varying stages of degradation, several types of Dominican dry forests can be identified, i.e., dense, open, and thorny dry forests. With increasing degrees of water scarcity and successive degradation, a decline has been noted in the amount of species, the average height of trees and the diameters of trunks. In contrast, increases were observed as with cactus, thornbush and grass species. In some areas, the dry forest may eventually thin out to semi desert wastelands (Bolay, 1997; Oehl-schläger, 1996; Deutsche Gesellschaft für Technische Zusammenarbeit, 1992).

Dominican Dry Forests as Livelihood Systems

Dry forests in the Dominican south-west provide the basis of livelihood for more than 20,000 woodworker families. Estimating that one extended family of this region includes approximately 7 persons (Erhart, 1995), every sixth inhabitant of the sparsely populated south-western region directly depends on natural resources for economic survival. While in 1993, there were 810,000 inhabitants in all of the south western provinces (Azua, San Juan, Elías Piña, Barahona, Baoruco, Independencia, and Pedernales), the share of dry forest inhabitants thus amounts in that region to nearly 20% (Deutsche Gesellschaft für Technische Zusammenarbeit, 1994).

By examining the access conditions to dry forest livelihood systems as important elements of survival, by further assessing the quality of natural resources on which they depend, and by finally outlining the political and economic status dry forest dweller have at a national level, it could be stated that dry forest inhabitants of the south west are exposed to severe existential risks. This shall be proved from some of the particular characteristics inherent in their system of livelihood, i.e., insecure access to land, small family incomes, dependency upon outsiders, exposure to ecological risks, and weak political status.

Access to Land

A special feature of the majority of dry forest inhabitants is that they do not own any land. In order to obtain access to it, woodworker families will squat unused land, partly being government-owned, partly being private holdings. In most cases, dry forest squatters are tolerated by landowners and are not required to pay tribute or rent. Nevertheless, dry forest inhabitants' access to land resources is extremely insecure as it depends almost entirely on the good will of others.

Access to Economic Resources

The production of charcoal is the most important economic activity of woodworker households. In most cases, however, the income generated is insufficient for their survival. Household incomes are, therefore, often composed of a variety of alternatives. Apart from making charcoal, many families produce timber, rear goats on natural forest pastures (both for home consumption and commercialisation), produce brooms and wooden handicraft, and collect medicinal plants and honey.

For many families further additional income is necessary to adequately support the household. One of the alternatives is seasonal work as farm labourer on private agricultural holdings in both the San Juan Valley and in the agrobusiness sector of Azua Valley. However, in view of the region's high level of unemployment, those jobs are insufficient in number to support all of the woodworkers' households requiring additional income.

In order to supplement their diet and relieve the family budget of the burden of grocery bills, woodworkers whose communities are situated close to areas of higher elevation with increased rainfall and non-irrigated agriculture will often maintain *conucos*, i.e., small slush and burn plots of about 1 ha where, for example, cassava, pigeon peas, plantains, pumpkins and papaya are cultivated.

The limited economic resources of dry forest households, usually located in small remote forest villages, are revealed by narrative patterns which woodworkers use when asked to relate their situation to financially stronger rural groups. These include descriptions of their houses which are built of clay walls (instead of block) and earthen floors and roofs of palm leaves or zinc sheets (instead of cement floors and ceilings). Normally, houses are not equipped with electricity and running water.

Dependencies

In the most recent past and with the assistance of foreign development agencies, dry forest inhabitants having organised themselves won access to national markets to sell their produce. Traditionally, however, their livelihood system was mainly characterised by strong dependency upon intermediaries controlling trade with dry forest products and achieving net profits of up to 100% (or even more).

National markets for timber and charcoal, such as Santo Domingo, are more than 200 km away from the producer community. This distance cannot be managed by individual woodworkers alone. Unlike intermediaries, dry forest inhabitants lack the appropriate logistical and financial means (such as lorries) to reach urban wholesalers.

Ecological Risks

The Dominican Dry Forest constitutes a fragile ecosystem with uncontrolled exploitation of its natural resources causing ecological degradation and thereby threatening the basis of livelihood of its inhabitants. Their survival in the long term depends above all on a sustainable management of wood resources.

Pressure on forest resources grows stronger whenever the producer population increases, or if returns from the sale of charcoal and timber are low and woodworkers are thus forced to produce more in order to survive. Severe ecological damage is also caused if dry forest management is carried out in highly degraded zones or with inadequate methods such as the exclusive use of 'green' trees (*madera verde*) instead of 'dead' ones (*madera muerta*), e.g., for timber or charcoal production.

In order to construct a charcoal kiln, wood is cut to manageable size and stacked tightly together in conic form. Grass, leaves and lichen are placed over the wood and completely covered with soil to control air filtration. The kiln is fired from a small opening at one side. Within a few days, the smouldering fire inside the kiln will transform the wood into charcoal. One cubic meter of wood produces 6 bags of charcoal with one bag containing about 45 kilograms of coal.

Political Status

Forestry legislation   Since mid-19th century, several governments in succession have intended to protect Dominican forest ecosystems through 'off limits regulations' (Bolay, 1997), however, disregarding forest resources as the basis of livelihood for thousands of families. In 1962, the military-controlled National Forest Administration (FORESTA) was created in order to protect the remaining forest resources and to supervise and regulate the trade and sale of timber or charcoal. Military personnel carried out a rigid forestry policy which was in accordance with the environmental and political philosophy of then conservative President Balaguer.

During his leadership (1988 to 1996), the prohibition against cutting of trees and burning charcoal was intensified through different presidential decrees. Actual conditions of the country's ecology, however, show that those strategies were ineffective since forestry laws were applied in an inconsequent manner. Although it was provided that violations should be severely punished, it proved difficult to enforce the law at local level where forestry officers' extremely low pay made them highly susceptible to corruption. Environmental protection executive staff also was strongly affiliated with the president's clientele by direct appointment through his office. Due to frequent changes of posts within this group, forest inhabitants were continually confronted with new policy-makers. In 1992, for example, the new director of FORESTA implemented a strict enforcement of forestry laws which included the persecution and arrest of charcoal makers using military force. Urban consumers were urged to forgo the use of charcoal.

Since 1996, with the election of President Fernández from the Liberal Party, environmental reforms have been an important part of the government's agenda. Strategies which combine forest conservation with rural development are being formulated in a new environmental master-plan. Yet, negative consequences of the past decades are still prevalent as could be seen from the lack of competent and decisive officials willing to take active responsibility for the country's environmental and forestry policy. Moreover, the environmental sector's limited financial budget as well as the tactics of centralised planning may be seen as obstacles for effective environmental reforms (Ulbert, 1999).

Land reforms Up to the present time, Dominican forestry policies have shown a general tendency to focus on the effects of deforestation rather than its causes and measures of prevention. Strategies are required which combat the problem at its roots, for example, the implementation of an effective land reform whereby landless wood squatters are provided with survival alternatives to charcoal-making.

In 1962, and following international pressures, the Dominican government became committed to land reform activities. The Dominican Agrarian Reform Institute (Instituto Agrario Dominicano, IAD) was founded and made responsible for the distribution of arable land to landless rural families. It was planned that this process should be combined with an agricultural extension service and infrastructural support (Instituto Agrario Dominicano, 1994). In reality, those services have not been very successful and the quantity of distributed land has not challenged the existing Dominican land tenure system which is characterised by the Latin-American *minifundia-latifundia* pattern. In the Dominican case, 2% of the population owns 54% of the arable land whereby 82% of the population is in possession of only 13% of the land (Bolay, 1997; Ferguson, 1993; Ceara, 1987).

Titles on Land The majority of rural groups who claim small pieces of land (*minifundistas*) do not possess titles of private land. However, if they can demonstrate that they have cultivated their land for about twenty years, in theory, they can claim customary right for it. In practice, however, it is quite an exceptional case that smallholders register their land and gain legal protection on it since the legal proceedings are both expensive and difficult to organise.

Reducing Risks: Dry Forest Management in the Community of Chalona

With the examination of some of the particular socio-economic and ecological characteristics of dry forest livelihood systems including their political status (towards the state), it was shown that woodworker families in the Dominican south-west only have a limited scope of action in order to secure their survival as their livelihood system is exposed to existential risks. However, given the fact that dry forest management is carried out by thousands of families, certain advantages inherent in this system of livelihood could be assumed that were able to buffer the mentioned risks. The advantages as well as further risk reducing strategies (either carried out by dry forest inhabitants themselves or by outsiders) are identified as follows using the case of charcoal burners in the community of Chalona.

The Community of Chalona

Research Area The results are based upon field research carried out in the community of Chalona of the south-western province of San Juan being located about 8 km south-west of the provincial capital San Juan. As could be seen from Figure 10.2, the Neiba mountain range marks the southern boundary of the high Valley of San Juan. During the first half of this century, the valley used to be a zone of extensive cattle raising. At present, due to an irrigation system installed at the beginning of the 1970s in the course of agrarian reform, intensive agriculture is carried out on the fertile soils of the valley bottom mainly in the form of cultivating red beans.

Chalona community extends over some 49 km². Several types of landscapes can be found depending on different levels of altitude, i.e., the arable valley bottom (at an altitude of about 400 m in the northern part of the community), the dry forest zone (at an altitude up to 500 m along the foothills of the Neiba mountains), and the region of humid mountain forest (at an altitude up to 1,500 m in the southern part of the community).

Characteristics of the Target Group The survey was carried out in two villages of the dry forest zone (Suarez, Los Cerros) which number together about 700 inhabitants. A total of 126 villagers were consulted by applying participatory research methods such as formal and informal group discussions, semi-structured interviews with key informants, participatory mapping and diagramming, and 'transect walks' together with members of the target group (Chambers, 1992).

Charcoal-burning is the villagers' most important economic activity. Additionally they rear goats and sometimes also produce timber. Some villagers cultivate small subsistence plots (*conucos*) at an altitude of about 800 m in the zone of humid mountain forests. Furthermore, the majority of the villagers interviewed mostly work seasonally on private agricultural holdings of the irrigated bottom part of the community. One of the places where this type of work is offered is the village of Los Tamarindos. Here, for reasons of comparison, some owners of medium sized farms, chiefly being beneficiaries of land reform activities, were included into the survey.

The communal provision of infrastructure turns out to be much more improved in the farming villages of the bottom part than in villages of the dry forest villages. For example, the latter can be reached only by steep paths whereas the villages on the valley bottom can be reached easily by small roads. The children of woodworker families have to take a three hours walk in order to reach secondary school whereas children of farmer families have to walk one hour with motorcycle taxis often providing a lift.

Charcoal: Offering a Product with a Secure Demand

The responses of Chalona's dry forest inhabitants made obvious that most of the families had originally come from communities close to the Haitian border. They had arrived in Chalona between 1930 and 1950. Both, the shortage and the poor quality of arable land in their mountain communities were mentioned as reasons for having left their places of birth. In Chalona, they not only had found space to set up new agricultural subsistence plots in the uncultivated mountain forests owned by the state, but, and above all, they had been attracted by the demand of wage labour required throughout the year in cattle raising farms in the Valley of San Juan. Hence, the possibility to produce charcoal from dry forest resources was not mentioned as a reason to settle down in Chalona.

Yet, with the agricultural transformation of the Valley of San Juan from a cattle raising zone into a region of irrigated farmland since 1970, less wage labour than before was required in the valley with remaining working opportunities showing a strong seasonal demand. Therefore, Chalona's dry forest inhabitants had to look for a regular alternative source of income. Like in many other dry forest communities of the south west at that time, the production of charcoal in Chalona as an alternative income strategy became increasingly significant.

The process has to be seen on the background of a growing national demand for charcoal, which is commonly consumed by poor urban groups. As a result of the country's beginning economic crisis, urban marginalisation processes became extraordinary noticeable in the 1970s. The decline of prices on the world market for traditional Dominican export goods in combination with growing foreign exchange debts had further forced the country to carry out harsh austerity measures under the assignment of standby agreements with the International Monetary Fund (IMF) during the beginning of the 1980s (Hellstern, 1993; Deere, 1990). In rural areas, the rising costs of living, wage cuts and high rates of unemployment resulted in extensive migration movements towards the urban sector, which in 1980 assembled already 50% of the country's population as compared to 60% in 1990 (Barrios and Suter, 1992).

In general, charcoal for cooking is mainly used by poor urban groups. Coal is given preference to propane gas which has comparatively higher costs of investment in the form of gas bottles and cooker. In contrast, stoves for charcoal are easy to build from waste-products of metal, and the fuel could be bought daily in small amounts, thus stressing less the household budget of poor groups than with monthly expenditures for a gas bottle filling (Erhart, 1995).

Hence, with the production of charcoal, a product of an extra-ordinarily secure demand is offered on the national markets. This demand might even increase in future taking into account the steadily growing number of poor urban groups.

Dry Forest Management: A Set of Flexible Work Activities

In order to secure their family income dry forest households combine forest management with wage labour and subsistence farming. Activities resulting from dry forest management such as the production of charcoal and timber and goat rearing are predestined for interlinkages between several income and work strategies carried out in a flexible way throughout the year.

In the community of Chalona, dry forest inhabitants practise wage labour on agricultural farms in the Valley of San Juan. As outlined before, the activity is limited to just a few weeks per year. As given in Table 10.1, work in the form of agricultural day labour is offered almost exclusively in November, i.e., preparation of fields to grow beans, and in February, i.e., harvest of the crop. Despite the limited period of time, the daily wage originating from this work turns out to be more lucrative than the proceeds from charcoal-burning. In 1995, for example, the eight hours working day of an agricultural labourer would earn him about one and a half times more money than an eight hours charcoal producing job.

Table 10.1 Yearly calendar of work activities of Chalona's dry forest inhabitants

	Wage labour	Dry forest management	Subsistence agriculture
January			x
February	x		
March		x	x
April		x	x
May		x	
June		x	
July		x	x
August		x	x
September		x	x
October		x	x
November	x		
December		x	X

In Table 10.1, it is further shown how Chalona's dry forest inhabitants coordinate the organisation of different work strategies during the year in relation to wage labour opportunities in the Valley of San Juan. Activities of dry forest management are practicable at any time during the year and, hence, they are most likely to be carried out in periods without wage labour. At that time, and apart from activities in the dry forest, some families carry out labour-intensive tasks on *conucos* such as clearing, burning and planting (January, March, April). Most of the crops are harvested during the second half of the year.

Producing Charcoal in Periods of Crisis

Above all, the production of charcoal could be seen to constitute an advantageous work strategy in order to cope with unpredictable economical stress such as costs of illness and debts (among others). For three reasons, it can be practised without long lasting periods of planning and preparation processes in advance. Frist, the knowledge of charcoal-burning is easily and quickly acquired. Secondly, investment costs required by the procedure are extremely low (simple tools, bags). And thirdly, working results are sure to prove after a few days, and not after months of risk, as in the case of agricultural cultivation, for example.

The production of charcoal represents an income strategy with lower risks involved during times of climatic shocks since it is rather independent of unfavourable weather periods such as frequent droughts. In the dry year of 1991, in Chalona for example, not only the harvest of the *conucos* had been affected, but also the yields of many agricultural holdings in the valley. For this reason the demand for wage labour in the valley declined and many dry forest inhabitants had to forego their income from farm work. The intensified production of dry forest products, however, helped to buffer their financial losses.

Availability of Intensive Family Labour

The security of the multi-compounded and labour intensive production system of dry forest inhabitants is guaranteed because each member of the extended families contributes with his or her working capacity. Households may also count on the help of their neighbours. Thus, with their high potential of labour, dry forest households possess an important 'net asset' in order to reduce risks and secure their survival.

Most of the work done by dry forest households is not determined by gender or age. In the community of Chalona, both men and women work as wage labourers in the valley and carry out tasks on their mountain plots. If a charcoal kiln is constructed, men are often supported by women and children. For example, they will help with the transport of wood to the kiln, with selecting charcoal out of the kiln and with packing charcoal into bags. The number of household members needed to manage on-going work finally depends upon the prevailing economic pressures perceived by the family.

In some particular spheres of responsibility, there exists a gender specific division of labour. Work which has to be accomplished with axes or machetes is normally carried out by men (preparing wood for kilns, clearing fields, producing timber), while the commercialisation of charcoal is primarily done by women – as it is the case with shopping and housekeeping. Women and small children are obliged to fetch water from remote springs. However, if the quota of daily work is compared between men and women, obviously, female members of Chalona's dry forest households turn out to be more heavily burdened than men.

Division of labour in Chalona exists in that dry forest households which belong to the same extended family, which share friends or are neighbours will help each other with the gathering of their goat herds in the evening, for example.

Strategies of Commercialising Charcoal

In the specific case of Chalona charcoal-burning is a main and profitable strategy because charcoal can be commercialised directly by the producers in the neighbouring City of San Juan with its about 60,000 inhabitants in 1993 (Deutsche Gesellschaft für Technische Zusammenarbeit, 1994). Thus, it is ensured that the investment of labour pays for itself within a short time. This has to be seen as an exception since for most of the south-western dry forest communities the market for charcoal is situated far away from the producer societies. Thus, with their strategy of direct commercialisation, Chalona's charcoal burners have never been dependent on scrupulous intermediaries, a fact which guaranteed them also higher returns from the sale of their product.

Twice a week, many women of Chalona's dry forest villages join and make their way down to the Valley of San Juan. Their walk starts before sunrise and lasts up to three hours. Charcoal is packed into bags and carried by donkeys. In San Juan, the women wander through poorer urban quarters extolling their product, with the sale of charcoal from house to house lasting up to four hours.

In spite of the need of time and physical strain, most of the women surveyed did not want to forgo their job of commercialising charcoal. As compared to women of other dry forest communities, they hold the returns from the sale 'in their own hands', and may thus decide on the type of investment (such as shopping weekly in San Juan after having finished work).

Proyecto Bosque Seco: Reducing Risks with Outside Assistance

The livelihood system of dry forest inhabitants, which at first sight seemed to be exposed to severe risks, turned out to be less vulnerable than expected. The main reason behind is that both existential advantages inherent in this system and risk reducing strategies applied by dry forest inhabitants themselves could be identified.

In the present analysis, however, the impact of ecological risk factors upon dry forest inhabitants (such as the threatening of the basis of their livelihood by the uncontrolled exploitation of forest resources) were not considered. Also, the question how dry forest inhabitants can cope with their insecure access to forest resources, though being the most important element for the security of their survival, has not been solved so far. The following chapter is meant to demonstrate that it has been possible to reduce those risk factors for many south-western dry forest inhabitants by measures of a German-Dominican development project run by the German Agency for Technical Cooperation (GTZ).

The project 'Rational Management of the Dry Forest' (*Proyecto Bosque Seco*) operates from the provincial capital Azua with the corresponding project area covering all dry forest communities of the Dominican south-west – see Figure 10.2. The primary targets of *Proyecto Bosque Seco* are twofold, i.e., the rational management of dry forest resources and the improvement of living conditions of forest inhabitants.

Organising Dry Forest Inhabitants

Promoting the organisational capacities of forest inhabitants represents a key strategy of *Proyecto Bosque Seco*. Up to present, 3,000 families out of 40 south-western dry forest communities have organised themselves in 81 grass roots organisations with the Azua based umbrella organisation (Federación de Productores y Productoras del Bosque Seco, FEPROBOSUR), having developed into an influential producer society being recognised both at the regional as well as national level. In the community of Chalona, most of the dry forest inhabitants are organised in two grass roots organisations affiliated to FEPROBOSUR.

Sustainable Dry Forest Management

At the beginning of the project in 1986 to 1989, it was assumed that dry forest resources in the Dominican south-west should be written off as being 'not anymore utilizable' because of the degenerated state of the dry forests. The forest producers should be convinced not to use the forest, which was seen as the only way to prevent further degradation. Hence, additional income strategies were developed like goatkeeping, producing wooden handicraft, collecting medicinal plants, etc. Unfortunately, those project activities did not arouse much interest among forest inhabitants who were worried that their primary economical basis of livelihood, i.e., the production of dry forest products, was put at risk. Therefore, in the second phase of the project in 1990 to 1992, the working concept was reorganised. Instead of the utopian restrictions put upon the usage of wood resources, a concept for a sustainable dry forest management was developed, aiming at the protection of dry forest resources without preventing its human inhabitants from using them. The approach was the result of a systematic forest inventory ('stocktaking') of the project area's remaining forest resources whereby the real potential of dry forests was calculated by making use of the woodworkers' knowledge. It turned out that the potential of this natural resource was more significant and generated more income for woodworker families than it had ever been assumed before (Deutsche Gesellschaft für Technische Zusammenarbeit, 1992).

The project's new concept of a rational forest management was based on easily comprehensive methods. For users, the immediate benefit had to be apparent. One of such methods is the exclusive use of 'dead' wood (*madera muerta*) instead of 'green' wood (*madera verde*). A further example is the elaboration and implementation of individual forest management plans for each community (Oehlschläger, 1996). Meanwhile, about 60,000 ha of dry forest are managed with ecologically and economically sustainable methods in the project area. The consultation of Chalona's dry forest inhabitants brought as result that the majority of people then accepted the use of 'dead' wood for the production of charcoal and timber.

Legal Protection for the Production of Dry Forest Products

At the level of government, negotiations were carried out by advisers from *Proyecto Bosque Seco* and FORESTA. As a result, official permits to produce and commercialise forest products are now given to organised wood worker families. In return, dry forest inhabitants are obliged to respect a specific quota system which regulates and restricts the quantity of forest products controlled by the forestry authorities.

Like in other dry forest communities organized under FEPROBOSUR, timber and charcoal no longer have to be produced in a quasi illegal way in Chalona. Arbitrary repressions by forestry officers against the economical activities of wood workers (such as the arrest of Chalona's women on their way to sell charcoal in 1992) belong to the past. Experiences as such were decisive in that the community's dry forest inhabitants organised quickly in producer associations and in that they joined FEPROBOSUR.

Property Rights on Dry Forest Land

An additional incentive for wood worker families to manage dry forest resources in a sustainable way constitutes the recent possibility to gain property rights on dry forest land. This arrangement is based upon negotiations between the project experts and the Dominican Agrarian Reform Institute (IAD). Associative land titles on unused and government-owned dry forest land are now given to the grass root organisations of forest inhabitants. From 1994 to 1998, 167,266 ha were distributed to organised wood worker families by IAD authorities. In the community of Chalona, both in the village of Los Cerros and Suarez, a total of 3,538 ha of dry forest land were given to the villagers' organisations.

Direct Commercialisation of Dry Forest Products

One of the primary responsibilities of FEPROBOSUR is the direct commercialisation of dry forest products both at local and national markets. By means of direct commercialisation, the organised dry forests inhabitants have a chance to make higher returns from the sale of their products than by commercialising them through intermediaries. Thus, they can accept to produce a comparatively smaller amount of dry forest products as before (under the regulation of the quota system) which of course has positive effects upon the regeneration of the forests.

The direct commercialisation of dry forest products is mainly possible, because members of the FEPROBOSUR committee gradually gained confidence to impose the producer society's interests on competitive groups both on the local and the national level, e.g., the group of influential intermediaries. In order to eliminate the smuggling with illegal forest products FEPROBOSUR manages a strategic check point on the single main road leading to the country's capital in co-operation with the National Forest Administration.

In Chalona, timber, posts and railway sleepers made from dead wood are commercialised on national markets with the support of FEPROBOSUR. The guaranteed additional returns from timber production prove positive effects on the household income of the community's woodworker families. However, and mainly due to the pressure exerted by Chalona women (unwilling to lose their field of economic operations), the commercialisation of charcoal continues to be carried out in the regional centre of San Juan.

Inside View on Survival Strategies

It has been pointed out that dry forest inhabitants in the Dominican south-west are in the position to reduce some of the risks inherent in their livelihood system quite independently with the process being supported by the German-Dominican development project *Proyecto Bosque Seco*. By getting organized in grass root organisations and in the producer society FEPROBOSUR, every tenth wood worker family of the region has by then got under the aegis of the project. The families involved both profit from their legally confirmed access to forest resources and the reduced ecological risk for the community through the application of ecologically sustainable methods of forest management. In many communities within the project area pressure on the remaining dry forests has declined visibly in just a few years with the fragile ecological situation having clearly been stabilised (Bachmann, 1997). Thus, a further indicator of social vulnerability, i.e., the limited recovery potential of an ecologically endangered survival system, proves negative.

After having identified factors which reduce the potential of vulnerability of Dominican dry forest inhabitants, the people affected will finally be given a chance to express their opinion. The consultation of Chalona dry forest inhabitants made obvious that, for themselves, they evaluate the situation differently.

Producing Charcoal: The View of the Charcoal Burners

Wood worker families of Chalona associate ambivalent feelings with their situation as charcoal burners. They appreciate the opportunity of producing charcoal on which their livelihood system largely depends, but at the same time, they complain about the physical workload related with this survival strategy. Men as well as women emphasize that charcoal burning represents a discriminatory activity of work. In the words of Edia, 43 years old, and Alfonso, 50 years old, this reads like:

> If you don't start off with a load of charcoal, there won't be anything to eat. We live on it. We are obliged to do it. / Charcoal is the last word you might utter in your life. It is the final word you can say. Of all jobs here in this country charcoal burning is the most intensive one. Nobody but us is so stupid as to work so hard on charcoal burning.

The women consulted point out that it is mainly them who face prejudices from other social groups when selling charcoal outside the community. Charcoal burners are said to be 'poor, unintelligent, backward and dirty', or in the words again of Edia:

> There are women in San Juan we should not approach (...). There are people in this town who maintain you shouldn't buy a single bucket of charcoal from dry forest women. They say you can't stand their unpleasant smell.

Discrimination of charcoal burners interviewed is also confirmed by some of their employers, i.e., farmers of the village of Los Tamarindos. They maintain that dry forest inhabitants 'are neither interested in education nor in development'. A young farmer of Los Tamarindos claims that this is the reason 'why for him as well as most of his neighbours it is unthinkable to marry a man or woman from the dry forest'.

Accordingly, some of the wood workers interviewed raise the point that in the Dominican society charcoal burners have the bad reputation 'of living from hand to mouth'. They are accused of 'appropriating somebody else's natural resources and exploiting them inconsiderately'. This type of prejudices makes it difficult for dry forest inhabitants to claim customary rights and titles on land. This is one of the reasons why Chalona's dry forest inhabitants do not possess property rights on high elevation areas of the community, although they have been using parts of it for decades under the mode of subsistence agriculture.

Feelings of Powerlessness

The marginal social status of dry forest inhabitants blocks them off when pushing interests to the level of national authorities. This might be illustrated by statements of interviewees acknowledging their own incapacities, i.e., lack of self-confidence, lack of information regarding their rights and legal proceedings, lack of personal contacts to influential persons, lack of time for unavoidable and frequent consultations of Santo Domingo City authorities, and lack of financial resources for the occurring costs such as lawyer's fees or bribes.

Chalona's charcoal burners are made aware of their lack of political influence as soon as they compare their situation with other social groups of the same community. On the one hand, Chalona inhabitants mention the case of wealthy and powerful San Juan cattle breeders who were capable of gaining property titles on land which they had acquired in the higher elevated mountainous region of the community within a short period of time. On the other hand, neighbouring farmers of Los Tamarindos were cited to have been given irrigated land due to the agrarian reform in the beginning of the 1970s. The interviewees claim that the favoured farmers were either members of the then ruling party or disposed of personal contacts to influential officers of the Dominican Agrarian Reform Institute (IAD).

At that time, land distribution in the Valley of San Juan had been carried out on the basis of a particular law within the Agrarian Reform Codex (*Ley de Quota Aparte*). The law requires that if an irrigation system is constructed by the government, a certain amount of the irrigated land is to be provided to some of the neighbouring landless families. One of the Chalona charcoal burners interviewed was given the advice by a social worker of *Proyecto Bosque Seco* to find out at IAD whether, on the basis of this particular law, they were also authorised to acquire some parcels of irrigated land. Because of their long-lasting experience of social powerlessness and their low expectations regarding governmental support, this idea, however, did not provoke any enthusiasm among the dry forest inhabitants.

Future Prospects of Charcoal Burners

When asked which factors they consider necessary to improve their situation in a decisive manner, the majority of Chalona's dry forest inhabitants mentioned

- access to irrigated land, i.e., either access to irrigated plots of land in the valley, or access to irrigation techniques (irrigation canal) which permit agricultural cultivation close to their villages, and
- implementation of infrastructural improvements in their villages, particularly the need for a secondary school.

Many of the wood workers surveyed expressed the desire of becoming entirely independent of the necessity of producing charcoal if there should ever emerge a secure alternative source of income in their community, as it was said by Alfonso, 50 years old:

Everybody here – and I do not exclude myself – is sick and tired of producing charcoal. We should try a bit harder and forget about charcoal burning (...). We ought to start a new life, a life that has nothing to do with charcoal burning.

Young people, in particular, consider the idea of abandoning the community and looking for a prospect of social promotion elsewhere.

How Charcoal Burners Define Vulnerability: An Interpretation

As the interviews with Chalona's dry forest inhabitants prove, the charcoal burners consider themselves to be rather vulnerable. This self-assessment could be explained in socio-cultural as well as in psychological terms. In the Dominican Republic, the charcoal burning class ranks lowest in the rural social order as given. The charcoal burners' minor social and legal status, and their feeling of inferiority resulting from this situation, constitutes a decisive source of vulnerability. Consequently, the development efforts as asked for by charcoal burning families primarily relate to their social advancement.

The feeling of personal inferiority is most clearly revealed by the fact that the interviewed charcoal burners would not even investigate chances of gaining irrigated land that possibly would be entitled to them in the Valley of San Juan. The availability of irrigation land, however, was mentioned by them as the most important aspect of improvement since it would allow them to run a commercial agricultural enterprise like their respected neighbours of Los Tamarindos. This, finally, would be the step to enter the class of agricultural smallholders resulting in a considerable social improvement.

It was further suggested by the inhabitants interviewed that the improvement of educational opportunities could be very much suitable in order to minimise vulnerability.

Conclusions

It has been shown that dry forest inhabitants in the southwest of the Dominican Republic are a less vulnerable social group than might be expected first when considering the apparent risks impacting upon their quality of life, i.e., insecure access to resources of production, economic poverty, and dependence. The elements of uncertainty turn out to be effectively checked by certain risk reducing effects. The latter comprise a secure national demand for charcoal (being the most important dry forest product), specific interconnections of various dry forest work activities, the chronological order in the course of the year, and the extensive availability of family labour (being a decisive 'net asset' which dry forest households possess).

A further point to be made here is that statements about the type (or nature) of vulnerability of dry forest inhabitants in the area must be carefully analysed. Not all of them are subject to identical dependence as, for example, the dependence on intermediate trade. Furthermore, the degree of vulnerability is obviously less severe among dry forest families being involved in developmental activities such as the *Proyecto Bosque Seco* with such families having access to important resources of production and arranging means of environmental damage control within their respective community.

The inside view upon livelihood systems by dry forest inhabitants reveals that Dominican charcoal burners consider themselves extremely vulnerable with particular reference to their social status of marginality. For that reason, the support of 'empowerment', i.e., the support of individual psychical strength, courage, and self-reliance concerning legal political rights, will be an essential key factor when coping with vulnerability of social groups in endangered ecosystems. With regard to this, the developmental strategies of *Proyecto Bosque Seco* are seen to constitute decisive means of progress, i.e., promotion of the organisational capacities of forest inhabitants and support of their producer society which has gained political influence at both the regional and national level.

References

Bachmann, Y. (1998), *Analyse des Wandels von Trockenwaldbeständen im Südwesten der Dominikanischen Republik auf der Basis multitemporaler Satellitenbildauswertung*, Master Thesis, Geographical Institute, University of Mainz.

Barrios, H. and Suter, J. (1992), 'Dominikanische Republik', in D. Nohlen and F. Nuscheler (eds), *Handbuch der Dritten Welt*, No.3, Dietz, Bonn, pp. 373-96.

Basilis, G. (1995), 'Presente y Futuro de los Bosques Dominicanos', in Fundación-Friedrich-Ebert (ed), *Agenda Ambiental Dominicana*, No. 2, Buho, Santo Domingo, pp. 1-15.

Bohle, H.G., Downing, T.E. and Watts, M.J. (1994), 'Climate Change and Social Vulnerability: Towards a Sociology and Geography of Food Insecurity', *Global Environmental Change*, vol. 4 (1), pp. 37-48.

Bolay, E. (1997), *The Dominican Republic: A Country between Rain Forest and Desert, Contributions to the Ecology of a Caribbean Island*, Margraf, Weikersheim.

Cassá, R. (1991), *Historia Social y Economica de la Republica Dominicana*, No.2, Alfa & Omega, Santo Domingo.

Ceara, I. A. de (1987), 'Land Tenure and Agroforestry in the Dominican Republic', in J. Raintree (ed), *Land, Trees and Tenure*, Nairobi, Madison, pp. 301-22.

Chambers, R. (1989), 'Vulnerability, Coping and Policy', *IDS Bulletin*, vol. 20 (2), pp. 1-8.

Chambers, R. (1992), *Rural Appraisal: Rapid, Relaxed and Participatory*, Institute of Development Studies Discussion Paper No. 311, Brighton.

Deere, C. D. (1990), *In the Shadows of the Sun: Caribbean Development Alternatives and U.S. Policy*, Westview Press, Boulder.

Deutsche Gesellschaft für Technische Zusammenarbeit (1992), *Programa de Acción de Bosque Seco (PAB)*, GTZ, Instituto para el Desarollo del Suroeste, Azua.

Deutsche Gesellschaft für Technische Zusammenarbeit (1994), *Lineamientos para un Plan Regional de Desarollo del Suroeste de la República Dominicana*, GTZ, Azua.

Erhart, M. (1995), *Tropenwaldschutz durch Bodenbesitzreform und nachhaltige Bewirtschaftung: Der Fall Dominikanische Republik*, Hochschulschriften No. 19, Metropolis, Marburg.

Ferguson, J. (1993), *Dominikanische Republik: Zwischen Slums und Touristendörfern*, Dipa, Frankfurt.

García, R. (1994), 'Diversidad, Endemismo y Especies Amenazadas en la Flora de la Isla Española', in Fundación-Friedrich-Ebert (ed), *Agenda Ambiental Dominicana No.1*, Buho, Santo Domingo, pp. 25-35.

Hellstern, E. (1993), *Soziale Differenzierung und Umweltzerstörung: Bäuerliches Wirtschaften in einer Bergregion der Domikanischen Republik*, Sozialwissenschaftliche Studien zu internationalen Problemen No. 184, Breitenbach, Saarbrücken, Fort Lauderdale.

Instituto Agrario Dominicano (1994), *La Reforma Agraria en el Caribe: Su Inserción en los Procesos de Cooperación e Integración Caribeños*, IAD, Santo Domingo.

Informe Nacional (1991), *Conferencia Mundial de las Naciones Unidas sobre Medio Ambiente y Desarollo, Brasil 1992*, República Dominicana, Santo Domingo.

May, T. (1997), 'Bergwälder in der Dominikanischen Republik: Ökologie, Nutzung und Schutz', *Geographische Rundschau*, vol. 49 (11), pp. 662-7.

Oehlschläger, C. (1996), 'Trockenwaldbewirtschaftung in der Dominikanischen Republik', *Forst und Holz*, vol. 51 (6), pp. 169-74.

Perelló-Aracena, F. (1991), *Energy: Environmental Long-term Strategies for the Dominican Republic: Development and Application of an LP-Model to the Country's specific Energy, Economic and Environmental Conditions*, Volkswirtschaftliche Beiträge No. 127, Schulz-Kirchner, Idstein.

Ulbert, V. (1999), *Partizipative Gender-Forschung: Umweltprobleme und Strategien der Ressourcennutzung in der Dominikanischen Republik*, Freiburg Studies in Development Geography No. 17, Verlag für Entwicklungspolitik, Saarbrücken.

11 Actors, Structures and Environments

A comparative and transdisciplinary view on regional case studies of global environmental change

GERHARD PETSCHEL-HELD, MATTHIAS K.B. LÜDEKE
AND FRITZ REUSSWIG

Introduction

Modern Times and Global Change

In former years, let's say at the time of Charlie Chaplin's famous movie 'Modern Times', the various contributions in this volume probably would have been located under a different headline, particularly the notion of 'changing environment' would not have been as dominant. Though many of the aspects reported, like different cultural habits, hardships of poor people trying to meet ends, or more or less ignorant governments failing to provide sufficient support for these people, might also have been elements of such a report. Many of the coping strategies presented in the contributions to this volume are traditionals which have existed for a long time such as the transhumant system of the Massai pastoralism in Tanzania (chapter 6) or the risk-handling system of Huancar pastoralists in Argentina (chapter 9).[1] On the other hand, these traditional risk handling systems are increasingly endangered due to a number of social, political and economical factors. Again, this is nothing new in principle. These kind of changes have occurred throughout history and are also described in a number of the case studies collected here. Examples are the Iloikop wars in Tanzania in the 1840s and 1850s (chapter 6) or the changes, now to be located in the past, due to Apartheid politics in South Africa (chapter 4). So what are the new and outstanding facets and constraints of the current trends in 'coping with changing environments'?

Phenomena of Global Change

The following list of recent developments and trends, taken from the annual reports of the German Scientific Adivsory Council on Global Change (WBGU), might help to give hints to answer the question of what type the novel trends are (WBGU, 1993, 1994, 1997, 1999 a, b, c):

- modification of the physico-chemical composition of the atmosphere and subsequent possible climate change (greenhouse effect);
- soil degradation of all types, e.g., wind and water erosion, compaction, acidification, etc.;
- reduction of natural ecosystems by area and quality, implying significant losses of biodiversity, e.g., deforestation;
- pollution of freshwater resources and coastal zones, e.g., by agricultural and/or industrial activities; and
- massive land-use and land-cover changes throughout the world with global relevance due to changes in the water and nutrient cycles.

These are all aspects of global environmental change (GEC) signifying that humankind has started to transform the face of the earth in an unprecedented way. Though environmental changes have always occurred (just refer to Plato's dialogue *Kritias* written some 2,400 years ago), these trends unveil new properties, both in terms of scale and of irreversibility. Consequently, local coping strategies in face of environmental risks have to be put into the context of these facets of GEC which itself is difficult to be defined in more detail.

Moreover, there are social, political, and in particular economic changes which, with respect to scale and irreversibility, exhibit similar features as the natural trends listed above and which are most often related to or consequence of these environmental changes. Examples include the increase of international agreements, e.g., the growing number of environment related conventions of the United Nations, the so-called globalization of the economy and of communication (world wide web) or the improvement of the medical sciences connected with a rapid and sustained population growth. There are many more of these social trends relating to Global Change (GC). Again, coping strategies are embedded into this process of Global Change and their effectiveness have to be reflected against them as follows:

to what extent do coping strategies itself augment Global Change processes? Candidates, for instance, are deforestation processes in the course of slash-and-burn agriculture reported from the Eastern Miombo Highlands (chapter 5), the Tanzanian Maasailand (chapter 6), the Atlantic rainforest of Southeast Brazil (chapter 8) or due to extraction of wood for charcoal production in the Dominican Republic (chapter 10);

how far are coping strategies already confronted with GC induced environmental and/or institutional risks? Here, most of the contributions are somehow faced with so-called cumulative elements of Global Change (Stern, Young and Druckman, 1992), e.g., soil erosion in almost each of the regions reported on, or, more specifically, decreasing agricultural yields in spite of increasing inputs (chapter 2); aspects of systemic changes, particularly climate change or changes in the world economic structure, can, for example, be found in chapter 5 concerning the cultivation of tobacco as a typical export oriented cash crop; and

how robust are the commonly adopted coping strategies described in the case studies against further, more drastic global changes? This question provides some background to all of the contributions and is explicitly addressed, e.g., in the introduction of chapter 9.

In this article, we will discuss these questions using the contributions of chapter 2 to 10 and the so-called syndrome concept as a tool for integrating, typifying and condensing case studies to a rather small number of archetypal patterns of Global Change (WBGU, 1993, 1994, 1997, 1999 a, b, c; Schellnhuber, Block, Cassel-Gintz, Kropp, Lammel, Lass, Lienenkamp, Loose, Lüdeke, Moldenhauer, Petschel-Held, Plöchl and Reusswig, 1997; Petschel-Held, Block, Cassel-Gintz, Kropp, Lüdeke, Moldenhauer, Reusswig and Schellnhuber, 1999). We will discuss in the next two sections actor related and environmental aspects of the syndromes and their relation to the regionally specific contributions to this volume. A more formal analysis using modern, qualitative modelling techniques will be performed before turning to the general question on the 'perspectives of smallholders in the age of Global Change'.

Syndromes as Regional Patterns of Global Change

It is often cited that Global Change is actually a regional or local change in the sense that it is the local action of people which govern the processes of the nature-humanity relationship (Kates, Turner and Clark, 1990). Yet, modern natural science has taught us that there are processes in the earth's atmosphere which transform these local actions into global reactions. Most prominent is the diffusion of greenhouse gases, in particular carbon dioxide, inducing a systemic global change of the atmosphere's energy balance. In the end, there might be a warmer world - at least in terms of global average temperatures. Yet, there is a high regional variability of global warming thus bringing back the effects of climatic change to the people in a variety of ways (Intergovernmental Panel on Climate Change, 1996). As an example, most climate models agree that there will be a pronounced increase in temperature in high latitudes, especially over land, whereas the warming is much more moderate in the tropics. In addition, there is a great regional variation in precipitation changes, both in sign and magnitude.

Besides these systemic changes (Stern, Young and Druckman, 1992), there is a number of so-called cumulative global changes, i.e., these phenomena receive their global relevance by occurring in different regions simultaneously. Examples include loss of biodiversity, soil degradation, population growth, the endangerment of food security, etc. It has to be noted that there are natural, i.e., in the current context environmental as well as social effects. Yet, it would be misleading to analyze these degradation processes in complete isolation from each other and from different regions. An obvious reason for this is that there are still global changes involved, including political as well as economic processes (see above).

Against this background we cannot separate between regional processes and Global Change. This raises the question, how to analyze Global Change without

- simplifying the unique features of local and/or regional situations; this particularly refers to the broad variety of risk handling strategies developed by local actors at different scales; the contributions to the current volume represent a small, but scintillating selection; for others see, e.g., Kasperson, Kasperson and Turner (1995) or Turner, Clark, Kates, Richards, Mathews and Meyer (1990); and
- paying too little attention to the interlinkages and similarities between regions and between processes, and thus to 'good' results from large-scale modelling and assessments; many modelling attempts of Global Change actually focus on these similarities, e.g., by specifying a single mechanism of land-use changes assumed to be valid throughout the world (Zuidema, van den Born, Kreileman and Alcamo, 1994; Alcamo, Kreileman, Krol, Leemans, Bollen, van

Minnen, Schaefer, Toet and de Vries, 1998) or throughout a single cultural perspective (Strengers, Elzen and Kösters, 1997).

So far, there are no convincing 'coping strategies' bridging these two extremes of global change research. This is also appreciated in the science and research plan of the International Geosphere-Biosphere and International Human Dimensions Programmes' core project 'Land-Use and Land-Cover Change' (LUCC) (Turner, Skole, Sanderson, Fischer, Fresco and Leemans, 1995, p. 12):[2]

> Modelling the dynamics of land-use and land-cover change has been hindered by large variations of those dynamics in different physical settings. Global aggregate assessments based on simple assumptions miss the target for large sections of the world, while local and regional assessments are too specific to be extrapolated to wider scales.

The approach which we want to present here tries to find a bridge between these two extremes by analyzing regional and local settings which, though geographically different, are similar in terms of

- *actors*, i.e., what social, economic, and political groups are involved in the process of land-use (or global) changes or, more specifically in the focus of the current volume, in the coping strategies in face of changing environments;
- *structures*, i.e., how do actors interact with each other and with their environment and what are the social frameworks for their strategies; and
- *environment*, i.e., what functions of the natural environment are of particular importance in this setting, e.g., is it purely the production function or is their some recreation function involved as, for example, in chapter 6 where tourist hunting activities are mentioned to be relevant for land-use conflicts in the Tanzanian Maasailand.

Though in the present article we mainly focus on the problem of land-use, the approach can be applied more generally, i.e., to Global Change in its entirety. This approach is similar to that taken by the field of pattern recognition known from modern natural sciences (Haken, 1983; Nicolis and Prigogine, 1977). We, therefore, call the similar settings obtained from the analysis *patterns* or *syndromes* of Global Change (see also chapter 1 for an overview of syndromes proposed at the time being).

Syndromes of Global Change: State of Affairs

In order to perform a pattern analysis of globally relevant interactions between nature and humankind, it is not possible to rely on a single best strategy, possibly based on a large set of data. The reason is that the facets of these interactions are so widespread that it is completely unclear which data, both from the natural as well as from the social sciences, are needed to bring about a reasonable and politically usable set of patterns. Also, experience with the most modern methods of pattern recognition has shown that these techniques might tell you something about the topology of the pattern space, i.e., how much, within which geometry and which functional neighborhoods, etc. The interpretation of the results, however, is rather difficult, hiding the actual mechanisms in a black box.

For these reasons, the syndrome approach tries to follow a more semantically oriented path of analysis, and we start from a catalogue of so-called *symptoms of Global Change*.[3] A symptom is already a generalized expression of qualitative and/or quantitative changes in the Earth System which are considered to be relevant for Global Change. 'Generalized' means that regionally specific processes and phenomena can be mapped onto one single symptom. As an example let us consider the symptom 'intensification of agriculture'.

In a number of studies, also from this current volume, the shortening of the period before slashing and burning primary or secondary forests in order to cultivate the area is alluded to as a possible way of preventing a decrease in agricultural yield in face of a degrading natural resource. It is reported, for example, about the Atlantic rainforest in Southeast Brazil (chapter 8) that

> (d)ue to the short period of settlement, no stable rotation system for farming has been established. (...) Sustainable land use either requires inputs of fertilizers (a not very realistic option due to the lack of capital found) or large allotments that allow for the inclusion of adequate fallow periods. Presently, land managers can still profit from soils and vegetation by slashing and burning the well developed forests.

In other cases, however, intensification might refer to an increased input in terms of implementing irrigation schemes. For example, in the Dominican dry forest (chapter 10) we can observe that, 'due to an irrigation system installed at the beginning of the 1970s in the course of agrarian reform, intensive agriculture is carried out'.

Currently, we are operating with about 80 symptoms from the natural as well as social sciences. The catalogue, however, is continuously modified and updated coinciding with the analysis of an increasing number of case studies. For an actual list of symptoms please refer to the latest annual report of the German Council on Global Change (WBGU, 1999 c).

Yet, Global Change is more than the pure addition or parallel occurrence of symptoms. It is the complex and sometimes highly inertial interaction between the symptoms which constitute the risks people are facing and have to handle. Taking the example of the Atlantic rainforest cited above (chapter 8), we see that there is already an indication for such an interaction, i.e., the 'soil mining character of agriculture' implying on the level of symptoms and their interactions that there is a relation between 'intensification of agriculture' and 'loss of soil fertility' or, even more general, 'soil degradation'.

Borrowing the idea from medicine, a syndrome is now defined as a typical cluster of symptoms and their interrelations. The notion of cluster refers to the concept that different syndromes only weakly interact, i.e., in order to describe, model, or analyze the basic properties of a single syndrome it is not necessary to take into account other syndromes explicitly. Rather they can be 'abstracted' in some weaker form (Petschel-Held, Block, Cassel-Gintz, Kropp, Lüdeke, Moldenhauer, Reusswig and Schellnhuber, 1999).

Case studies like those contained in this volume embody the basic information base used to specify the syndromes. They help to answer questions of what are the interactions between which symptoms. The patterns are formulated in an inductive step, i.e., it is not possible to give a complete set of criteria representing a rationale behind the formulation of a specific set of syndromes. It is therefore necessary to check for the explanatory power of this set and the details in terms of symptoms and interactions. Besides the obvious way of weak testing by discussing it with experts and sustained cross-check with other case studies, the ability to qualitatively reconstruct the time development described in the studies is of particular importance - for the concept of qualitative trajectories see also Kasperson, Kasperson, Turner, Dow and Meyer (1995). In order to do so in a formally consistent and stringent manner we make use of the concept of qualitative differential equations (Kuipers, 1994). This concept allows all time behaviors which are compatible with a qualitative description of mechanisms like 'A enforces B' or 'C decreases with D' to be determined. If the interactions between symptoms within a single syndrome can at least be qualified in terms like these, we obtain a variety of possible time developments to be compared to those found in the studies.

Table 11.1 Mapping between regional case studies and syndromes

Number of chapter and region of contribution	Syndromes involved
2 Upland area in Nepal	SAHEL
3 Laotian Forest	SAHEL OVEREXPLOITATION ARAL SEA
4 Cape Town, South Africa	SAHEL FAVELA
5 Miombo Highlands, East Africa	SAHEL DUST BOWL OVEREXPLOITATION
6 Tanzanian Maasailand	SAHEL DUST BOWL MASS TOURISM
7 Rural Botswana	SAHEL ASIAN TIGER
8 Atlantic rainforest, Southeast Brazil	SAHEL
9 Andean Mountains, Northern Argentina	SAHEL
10 Dominican Republic, Southwest	SAHEL OVEREXPLOITATION

Syndrome Guesses

Parallel with the more inductive approach to syndrome analysis, we start with a preliminary assignment of the case studies to the syndrome matrix as presented in Table 11.1. This assignment is based solely on the 'reading experience' of the various contributions and therefore helpful to provide a discussion guideline what is a particular strength of the syndrome concept in terms of its capacity to outline global environmental problems.

It is no surprise that the Sahel Syndrome occurs within all settings. The syndrome exactly deals with the different coping strategies of smallholders within marginal natural landscapes. In particular, the syndrome seeks to specify those strategies within which the network of feedbacks and interrelations finally endanger the livelihood of local people. In many cases, emigration is the only option. A paradigm for this syndrome can be observed in the former homelands of South Africa (chapter 4, p. 113):[4]

In the rural areas of former Transkei and Ciskei the livelihood was and still is based on subsistence agriculture. However, the subsistence basis has been severely damaged due to food crop cultivation on unsuitable land and due to overgrazing. This resulted in ongoing erosion processes, thus, further destructing the subsistence basis.

In the remaining sections we scrutinize in more detail the Sahel Syndrome. The remainder of this section sheds some light on the other syndromes tabulated above.

Overexploitation Syndrome This syndrome delineates the natural and social processes governing the overexploitation of natural ecosystems in the course of extracting biological resources (WBGU, 1999 c). Important social driving forces are present, in particular massive policy failure (corruption, the inability to enforce rules and regulations for sustainable use of biological resources, lobbyism, etc.). An important feature is the 'stabilization' of policy failure by successful overexploitation. Most prominent examples of the syndrome are sustained deforestation due to industrial logging activities and the boundless depletion of world-wide fisheries. Thus, neither the wood-gathering activities of smallholders nor the conversion of forests due to shifting cultivation are subsumed under the notion of the Overexploitation Syndrome what is the reason why the syndrome is only tabulated for the Laotian forest (chapter 3) and the miombo highlands (chapter 5).

Of particular relevance is the opening of forests by timber extraction activities, often encouraging massive settlement activities in its wake. An example can be found in the study on Laos (chapter 3) stating that 'due to the construction of weather tracks to transport timber, many formerly inaccessible villages had been connected to the outer world'.

Aral Sea Syndrome A discernible paradigm of this pattern is the ecological catastrophe of the Aral Sea caused by large scale water diversion for irrigation schemes in the now Caucasian steppe. The syndrome describes the ecological and social deterioration as a consequence of large-scale infrastructure projects (WBGU, 1999 a). In terms of global relevance, it is the erection of a total of over 40,000 large dams with a height of more than 30 metres which bestows to this pattern a global relevance (International Commission on Large Dams, 1984, 1988; McCully, 1996). So far, one river system which still exhibits a large untapped potential for hydropower is the Mekong which is one of the subjects dealt with in the case study on Laos (chapter 3). Besides the direct environmental effects of the construction process and the loss of ecosystems due to flooding and change of the physico-chemical regime of the river flow, it is the change of land-use rights and necessary resettlement programs which affect smallholders. On Laos, for example, it is reported (chapter 3) that there is 'evidence of various harmful and project-related effects such as declines in fish catches downstream of Nam-Kading-River, of some villages being impacted by the loss of riverside vegetable gardens and of increasing transport difficulties'.

Favela Syndrome The name *favela* stems from the Portuguese expression for informal settlements hinting to the defining character of the syndrome, i.e., environmental and social deterioration in the course of unregulated urbanization processes. Thus, the aspects discussed in the study on informal settlements in Cape Town (chapter 4) are by definition elements of this syndrome. The mechanisms described in the contribution point out an important property of the syndrome concept in that, as in medicine, one syndrome can vitiate another. Migration as a coping strategy against increasing impoverishment in rural areas can finally lead to environmental crises within settlement areas as such (WBGU, 1999 a).

Dust Bowl Syndrome First, this pattern of Global Change investigates the structures and processes associated with environmental and social quashing due to capital intensive agriculture, often cultivating cash-crops in developing countries, as described for the Tanzanian Maasailand (chapter 6) where, at present, 'about 7,000 ha are cultivated each year for the production of seed-beans, beans, wheat and maize' and where a 'severe consequence is the inaccessibility of important seasonal and farm-based water ponds' (for the Maasai herders).

Second, however, an important variation of this syndrome concerns plantations for wood production presenting *inter alia* an important source of wood for fire- and flue-cured tobacco production in the miombo highlands of Tanzania and Malawi (chapter 5). Here, the production of tobacco also falls under this syndrome.

Mass Tourism Syndrome The still increasing world-wide numbers of travel activities prompt massive environmental problems from over-proportionally high water withdrawals, sealing of soils for touristic infrastructure to social effects like in Tanzania (chapter 6, p. 169) where

> local Maasai communities hardly benefit from tourism related activities. The villages' shares from 'animal head fees' derived from hunting were not distributed by district authorities.

This exhibits an often found property of syndromes of Global Change, i.e., positively intended activities heading for the improvement of the livelihood of the local communities (e.g., by providing energy, income or water) bring about a number of negative side effects for the environment as well as for marginalized social groups. This might result in improvements of the human well-being and standard of living in spite of serious environmental or ecological degradation (Kasperson, Kasperson, Turner, Dow and Meyer, 1995). At the end, however, the degradation can even outweigh the positive effects and end in a massive deterioration of the living conditions of the local people, leaving them worse off than in the beginning.

Case Study Analyses: Actors and Environment

In general, all the case studies given in chapters 2 to 10 of the volume consider coping strategies of smallholder households in developing countries. This gave rise to the assessment which indicates the dominant role of the Sahel Syndrome. Thus, the question on converging or diverging traits of these strategies will be analyzed in more detail later. With respect to the extent to which other social players are involved and, thus, to what extent external dependencies do exist, the case studies vary. Yet, the main focus is on the role of the state and to a lesser extent on the influence and capacity of national and/or international companies.

State and Regional Governments

The state as a major actor relevant for the coping strategies of smallholders is evaluated in an ambivalent manner. In Laos, for example, the development of roads and infrastructure as well as the provision of logging licenses has some major direct negative impacts on the environment, i.e., deforestation. Yet, it is observed (chapter 3, p. 89, 92) that

> (t)he opening of forests by logging companies in the 1980s led to a rapid increase of paddy fields. (...) The lack of paddy fields (...) are identified to be the main reasons behind food security.

Table 11.2 summarizes the roles of the state as alluded to in the different studies. First of all, it should be noted that the state plays a significant role in all the regions investigated. Yet, the effects of state interventions vary greatly leading to quite different effects. Thus, though being important, it is not sufficient to investigate the state alone when addressing the three major questions raised above.

Table 11.2 Summarization of the roles of governments as actors impacting upon coping strategies of smallholders

Number of chapter and region of contribution	Role and influence of state or provincial activities with respect to smallholders' coping strategies
2 Upland area in Nepal	Prosecution of illegal timber cutting and/or alcohol production, both representing major (direct or indirect) sources of income.
3 Laotian Forest	Opening the country to foreign investors; road construction providing market access and/or cultivation of paddy rice; prohibition of trade of wood and wildlife products.
4 Cape Town, South Africa	Apartheid politics with social/economic/juridical marginalization of the African majority enforcing use of unsuitable land; provision of area and sometimes infrastructure in informal settlements
5 Miombo Highlands, East Africa	Rural political economy determines *inter alia* the size of the farms and the reliance of the local economy on exporting tobacco.
6 Tanzanian Maasailand	Major impacts on land-use rights; changes of agricultural structure towards cash-crop farming; ineffective social decision-making process by mixing traditional with modern institutions.
7 Rural Botswana	Highly effective governmental drought relief measures; unclear side effects of the measures, e.g., in terms of motivation for sustaining agricultural activities.
8 Atlantic rainforest, Southeast Brazil	Ineffective application of protection rules and control within reservation areas; more or less tolerated squattering and land-use by smallholders, including slash-and-burn techniques on relatively small plots.
9 Andean Mountains, Northern Argentina	State is seen as part of '*gobierno*' by the local people (Huancar), i.e., institutional risk; '*gobierno*' tries to introduce modern specialization instead of the more traditional system of risk aversion by diversification.
10 Dominican Republic, Southwest	State tolerates collection of charcoal by woodworkers; history of failing land reforms and forestry management plans due to corruption; positive examples in the particular study region under the NGO-participation of FEBROBOSUR.

Large Scale Companies

In some studies, the role of large, national and/or international companies or landowners is discussed. In general this group can be seen as a further player in the game of diverging interests. Often the state acts in favour of these large companies, e.g., in hope of some tax, license or exchange revenues. Again, the repercussion of this involvement is ambivalent as, e.g., on the one hand these companies act as employers thus providing labor and therefore income for the smallholder reducing their risk by diversification (see below). On the other hand, however, there can be significant marginalization of smallholders due to restriction of land-use rights by large-scale landholders.

The role of large-scale companies is more explicitly addressed in chapter 3 (Laos) and 5 (miombo highlands). For Laos, particular focus is put on the opening up of forests due to logging or hydropower installation activities with direct negative impacts on the environment and indirect effects for emplacing lowland rice paddies, thus providing fairly good income for the local peasants. In contrast, the role of large-scale companies in the miombo highlands is twofold (chapter 5, p. 144):

> Rent-seeking by European farm lobbies had long been prevalent in ... parts of Africa such as in Malawi. Aimed at reducing systematically the profitability of small farm cultivation through means such as unequal land distribution and contract farming, more of the smallholders (in Malawi) start now to realize their increased income generation possibilities by entering tobacco market relations after national tobacco monopolies and market access have been deregulated.

In summary, both cases indicate an ambivalence of large company actions as they pave the way for new income options for smallholders. Yet along this way, there had been massive environmental and institutional side-effects.

In the case of Laos, also the important role of international financing institutions such as the Asian Development Bank is addressed. These institutions are of particular relevance for the development of large-scale infrastructure programs like dams or roads. In other words, they are major actors within the Aral Sea Syndrome, as already mentioned above.

Environments

All of the coping strategies discussed in the various contributions depend to a certain extent on the opportunity for agricultural activities and for collection of natural products. There is hardly any capital available to purchase artificial fertilizer or to invest autonomously in irrigation schemes. Therefore, the agricultural yield (and to a similar extent the income from herding activities) strongly depends on the natural conditions, i.e. climate, soil, erosivity, surface water availability, etc. This section will show that all the regions considered can be classified as 'marginal'. This implies, among others, that any agricultural activity bears a great risk in terms of environmental degradation.

According to the information contained within the different papers, the climatic settings of the regions considered in chapters 2 to 10 range from mountain climates (Nepal, Andean), dryland regions (Tanzania, Malawi, Dominican Republic, Laos), arid to semi-arid conditions (Botswana) to humid or sub-humid climates (Laos, Southeast Brazil, Eastern Cape Province of South Africa).

There is, however, no general stringent information on the natural conditions for agriculture in the different regions. We therefore rely on a global assessment of marginality for agricultural land-use which is based on global data-sets of climate (temperature, precipitation, solar radiation, inter-annual variability of the growing season), soil properties (water retention, fertility), orography, and surface water availability (Cassel-Gintz, Lüdeke, Petschel-Held, Reusswig, Plöchl, Lammel and Schellnhuber, 1997; Lüdeke, Moldenhauer and Petschel-Held, 1999). The analysis is set-up as an expert system, i.e., the evaluation of marginality mimics an expert's argument using basic logical arguments like 'if soil fertility is low then growth conditions are bad'. The result of the analysis between 0 (no marginality) and 1 (highly marginal) can be depicted as a world map of agricultural marginality. It represents the natural component of the *disposition* for the Sahel Syndrome and indicates those regions whose natural prerequisites put a high risk on agricultural activity. A world map where regions particularly marginal are depicted dark is provided in Figure 11.1, having a spatial resolution of 0.5° x 0.5° which is 55 km x 55 km around the equator.

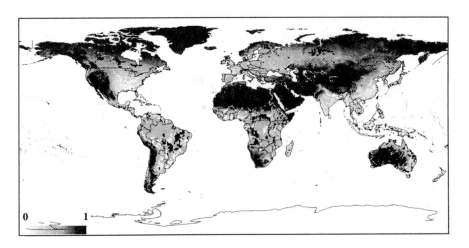

Figure 11.1 Global assessment of marginality with respect to agricultural activity

We have extracted the marginality values for the different study regions of this volume. The results, as presented in Table 11.3, suggest that all study regions show some degree of marginality. We also have included the result from the sensitivity analysis with respect to climate (Lüdeke, Moldenhauer and Petschel-Held, 1999).[5] This measures whether climate change will change the marginality in the respective region. As can be seen, the marginality in most regions depicts some degree of sensitivity which can augment the endangerment of the regional ecosystem by any agricultural or gathering activity. It has to be noted, however, that this measure only allows to speak about a chance of improvement or worsening and not about the actual direction. The reason is that we do not specify a certain direction of climate change due to the large degree of uncertainty in the model results.

It can be seen that the two mountainous regions, i.e., central Nepal and Huancar District in the Andes of Argentina have the highest marginality values of all study areas. This is mainly due to the rather cold climate which limits natural growth. This kind of limitation is reflected in the preference of herding activities as *one* major element of local coping strategies (see below). Yet, the sensitivity assessment indicates in both regions a slight chance for improvement, as the temperature limitation might be weakened due to global warming.

Table 11.3 Marginality M ∈ [0,1] of the study regions as obtained from the global assessment of agricultural marginality

Number of chapter and region of contribution	Value of marginality	Sensitivity against climate change
2 Upland area in Nepal	0.9 - 1.0	low
3 Laotian Forest	0.3 - 0.5	medium
4 Cape Town, South Africa	0.2 - 0.8*	low
5 Miombo Highlands, East Africa	0.3 - 0.85	medium - high
6 Tanzanian Maasailand	0.4	high
7 Rural Botswana	0.5 - 1.0	low - high
8 Atlantic rainforest, Southeast Brazil	0.05 - 0.25	none
9 Andean Mountains, N-Argentina	0.9	low
10 Dominican Republic, Southwest	0.2 - 1.0	low - medium

* Indicator for regions from which people in the informal settlements of Cape Town originally came from (Eastern Cape Province).

Some parts of the rural areas of Botswana have also quite high values of marginality. Here, water is the limiting factor for agriculture except for the Northwest where surface water of the Okawango reduces marginality. Care has to be taken, however, as the assessment does only include environmental damages due to agriculture. Therefore, damages due to large scale water diversion systems as sometimes suggested for the Okawango are not included. The environmental threats of such projects are investigated within the analysis of the Aral Sea Syndrome. Concerning sensitivity, it can be stated that for Botswana it shows a high spatial variability the pattern of which, however, does not coincide with the spatial pattern of current marginality. Therefore, it can be expected that there are winners and losers of climate change in this country.

Intermediate values of marginality are found for the three study regions in Africa (i.e., Eastern Cape Province of South Africa, Malawi and Tanzania), for Laos and for the Dominican Republic. Here, it is a mixture of climatic and soil properties which determine their values. Though there is a risk of soil degradation, it equally exists a fairly good chance for sustainable agriculture with little effort. This can, for example, be seen in the Laotian case of lowland paddy rice fields (chapter 3, p. 90) as

> the risks of soil erosion are reduced in the case of lowland rice cultivation. Another advantage of lowland rice cultivation is the long-term conservation of soil fertility. In Sangthong District, buffaloes and cattle graze on paddy grounds after the harvest, at the same time maturing the paddy fields.

The sensitivity analysis shows varying results for these regions mainly within an medium range. Thus, there is danger that in these regions a significant increase of marginality might occur, thus, sometimes augmenting the local dynamics of soil degradation and resource depletion. If all the study regions collected in the volume are to be compared, these regions can be considered as the most endangered ones in face of a changing world climate.

Having a closer look at the case studies, however, reveals an important deficiency of global analyses. As the assessments already aggregate, e.g., for soil fertility a spatial resolution of 1° x 1° degree is used based on the soil map of FAO (Zobler, 1986), any displacement processes to allotments of low productivity within a region of fairly good conditions is not sufficiently taken into account. Description of such kind of processes can be found, e.g., for the Eastern Cape Province (South Africa) or for the Atlantic rainforest (in Southeast Brazil).

The lowest marginality values can be found in the Atlantic rainforest in Southeast Brazil expressing rather favourable conditions except for low soil fertility (see above). This is also the reason why sensitivity in this region vanishes.

In summary, it could be stated that in all of the areas under study the medium to high degrees of ecological risks, together with the highly ambivalent role of state and business interventions, result in a similar pattern of ecosystem endangerment that is largely shaped by institutional factors. This situation requires highly adapted social strategies. A comparison and integration of the different strategies reported in this volume is given in the next section.

Comparing Smallholders' Coping Strategies

Here and in the following sections we want to talk about strategy which for itself is a dynamic notion. In principle, a strategy can be interpreted as a skill to aid survival in dangerous situation. The military usage of the notion of strategy is connected with a set of stratagem, or in other words a tool box of options which are used in the course of time to achieve a certain target. Transferred to the problem of coping strategies in changing environments, we, first, have to identify the toolbox of options that smallholders are using in different regions.

A detailed study of the contributions collected in this and other volumes (Little and Horowitz, 1987; Blaikie and Brookfield, 1987) reveal that, in principle, there are three main options for supporting livelihood:

- agricultural activity, i.e., growing crops or herding animals; the products can either be used for subsistence (LAS) or sold or bartered on markets (LAM);
- gathering or hunting pursuits, e.g., wood or non-wood forest products, game, which again might either be used within the own household (GS), sold directly, or sold after some value-adding production process (GM); and
- wage labor, often related to agricultural activities (LW), e.g., on large estates or holiday farms as mentioned in the study on the Atlantic rainforest in Southeast Brazil.

Some of the studies in this volume go beyond this set of options. In chapter 1, for example, it is noted that 'another coping mechanism being forced upon the people is to consume less than the minimum requirement needed for an active and healthy life'. And, similarly, in chapter 4 the question is raised whether rural-urban migration can be viewed as a coping strategy. In the following, we want to concentrate on the supporting options tabulated above. We refer to this set of options simply by the notion of 'option-box'.

The question we would like to address is whether the different strategies of all the smallholders reported in the various contributions obey a general rule or set of rules. This general rule would represent a major element of the mechanism of the Sahel Syndrome, particularly with respect to its social component. The specific strategy within a region would then be the result of combining this general rule with the situation and environment of the local people. Some aspects of this determining situation are actually outcomes of the local interaction between humans and nature, itself determined by previous decisions and actions. Thus, there is a significant degree of feedback in the system. This implies that the metamorphosis of the coping strategies to a certain extent are self-determined. We will discuss the implications of this feedback net and the degree of self-determinism by using a qualitative model of the smallholders activities in the next section.

Within this approach, major external elements play an important role, e.g., the influence of the state, on the one hand, and the environmental conditions on the other hand (with both alluded to in the previous section). For the moment, we want to assume that the environmental conditions as assessed and mapped in Figure 11.1 remain constant. Therefore, we can assume that the change in the natural conditions for agriculture such as soil fertility are solely governed by the local environmental effects in course of agricultural and/or gathering activities. Basically, this is one of the major feedback mechanisms within the study regions. As all the environments have to be considered as vulnerable, inappropriate strategies of resource use will finally lead to massive degradation and thus losses of the income base. Again, we refer to the modelling section below for the various implications.

The governmental ventures, as listed in Table 11.2, will be taken into account exogeneously when checking the explanatory power of the basic rules of the Sahel Syndrome to be delineated below. Before doing that, we first analyze the case studies with respect to the basic set of options listed above.

Most of the contributions in the volume report on the metamorphosis of coping strategies within the last years (even up to decades) and not on current strategies alone. These metamorphoses are summarized in Table 11.4. The attempt was made to depict not only the options actually used in the specific regions but also their current direction of change (to the extent it is known).[6] An increase in activity is indicated by an upward arrow ($\uparrow$), a decrease by the corresponding downward sign ($\downarrow$), and no change by a small circle ($\circ$). If no information on the direction of change is available, the corresponding box is simply checked by an x. The abbreviations used for the different labor activities are those introduced in the listing of the option-box above.

It is obvious that by forcing the different strategies into the rather simple scheme of five basic options, other basic social elements of the strategies are ignored. The scheme contains no information, e.g., on changes due to individualization in course economic development, as reported for rural Botswana (chapter 7, p. 181-2) since

> (d)rought was looked upon as a group problem which was to be solved by concensus within the group. Today, given the quickly modernizing Botswana society (...) drought is often perceived as an individual problem, with strategies taken up by individuals as outcomes of individually satisfying mediation.

Table 11.4 Generalized proximate coping strategies and environmental degradation (ED)

Number of chapter and region of contribution	Coping strategies [a]					
	LAS	LAM	GS	GM	LW	ED [g]
2 Upland area in Nepal	↓			X	↑	↑
3 Laotian Forest [b]	(↑,↓)	(↓,↑)	X	X		↑
4 Cape Town, South Africa [c]	↑				↑	↑
5 Eastern Miombo Highlands [d]	↓	↑		↑		↑
6 Tanzanian Maasailand [e]	(-,↓)	(↑,-)		X		(↑,°)
7 Rural Botswana	↓	↓			↑	X
8 Atlantic rainforest [a]	(↑,↓)				(0,↑)	↑
9 Huancar, Andean Mountains	X	X	X	X	X	
10 Dominican Republic [f]	X		X	(↑,°)	X	(↑,°)

[a] See text (p. 272) for an explanation of the abbreviations used.

[b] There is some difference in the allocation between LAS and LAM between the two major study regions of the contributions. Therefore two entries are made to indicate the differences.

[c] See Table 11.3.

[d] Here wood is not sold directly but used in the tobacco curing process.

[e] If there is a difference between the activities of farmers and of herders indicated in the text, the first entry relates to the new farmers and the second to traditional herders. The entry '-' means an absence of the respective pursuit.

[f] The second entries concern the values after implementing the *Proyecto Bosque Seco* in some dry forest communities.

[g] ED = environmental degradation.

In analogy to Stern, Young and Druckman (1992), we might therefore interpret the strategies as designated by the five basic elements of the options-box as 'proximate' strategies in contrast to a detailed sociological decomposition.

In addition, Table 11.4 also contains information on whether there is massive environmental degradation going on in the respective region (column ED) which is related to activities within people's livelihood. The scheme is analogous to the one for coping strategies. A question mark is put when no information is available. In summary, the result of the case study analysis as presented in the table represents a first step in the generalization scheme discussed above, i.e., general variables able to subsume details of the case studies are formulated and values assigned.

First of all, it is observed that the chapters 2 to 10 cover a broad range of different coping strategies, in particular when looking on current changes in composition. Basically, all strategies can be typified as *risk management by diversification* which fits well into traditional theoretical frameworks and is explicitly addressed in chapter 4 making the point that the 'more diversified the portfolio of strategies is a person or group can dispose of to cope with vulnerability, the better are the chances to keep up the *status quo* or even enhance the respective living conditions'.

The effect of non-available diversity can be seen, e.g., in the contribution on South Africa in that the almost purely subsistence based livelihood finally leads to migration as the only 'way out' from the degradation-impoverishment-spiral - what is a paradigmatic course of the Sahel Syndrome.

It also can be seen that from almost all regions a severe environmental degradation is reported. In particular this refers to high levels of different types of soil degradation, e.g. nutrient depletion or erosion.

We can now specify our basic question in more detail as follows. Is there a common set of rules which implies that the nine strategies shown in Table 11.4 will ensure, at least to some extent, the successful coping (or survival, even) of smallholders in spite of a degrading environment and the impact of governmental actors?

A Qualitative Model of Smallholders' Coping Strategies

In order to answer this question a qualitative modelling approach is applied in this section. It will be shown that this is possible to some degree by implementing a relatively simple cause-effect-net which connects several aggregated variables (Figure 11.2). This net is a specified part of the more general Sahel Syndrome which has been shown to cover the most important viewpoints of the case studies presented here. Taking the flow diagram of Figure 11.2, it becomes impossible to deduce the possible developments relying only on plausibility. That is the reason why we use the concept of qualitative differential equations (QDE). This methodology allows us to deduce *all* qualitative time developments of the considered variables which are in accordance with the cause effect net.

The method does not require further specification of the qualitative relations (e.g., by quantification) and, thus, allows a more direct mapping and evaluation of 'word-models' to be made. It may, therefore, fill the gap between traditional mathematical modelling and plausible reasoning, keeping (in part) both the exactness and clear distinction between assumptions and deduced conclusions of the former and the generality and the necessarily wide concepts of the latter.

The QDE concept formalizes the connection between cause-effect assumptions and the (endogenous) time development in form of qualitative trajectories. We, thus, can be more specific in our basic question to the following. Does the QDE-based qualitative modelling of the simple network of interrelations in Figure 11.2 reconstruct the qualitative observations and is, thus, 'non-falsified'.

Livelihoods of Smallholders from a Systemic Point of View

As a first step in classifying the livelihood resources of smallholders in developing countries (with respect to cybernetic aspects), we distinguish the activities which (a) directly rely on the use and management of natural resources and (b) activities which do not imply disposal or direct use of natural resources. In category (a) we find the different forms of farming (sedentary, slash and burn, livestock, etc.) including gathering (e.g., fuelwood) and hunting, while category (b) includes gaining income through wage employment in villages (e.g., in agriculture) and in town (e.g., in the service sector).

Now, there is an inherent ambivalence based in the structure of the activities of category (a) which can be elucidated by introducing three aggregated terms:

- QR, signifying the quality and quantity of the natural resource base important for smallholder production (e.g., soil for agriculture, productivity of grasslands for livestock or of woody vegetation for gathering fuelwood or exploitation of timber);
- LA, signifying the investment of labor into the activity (e.g., ploughing, weeding, gathering); this means that the important sub-allocation into the different aspects of agricultural work, especially the distinction between resource quality preservation measures and activities mainly aimed at short-term oriented exploitation is aggregated in one single term; compared to the more detailed

option-box described above, LA subsumes the categories LAS, LAM, GS and GM; and

- Y, signifying the physical output of the activity (crops, milk, fuelwood, etc.).

Obviously, the yield (Y) depends both on the quality of the resource (QR) and the investment of labor (LA). The specific form of this relation will vary largely over the different agricultural systems, including climate and soil conditions, techniques applied, etc. The properties which are common, however, are as follows. First, without any labor (LA = 0) or with vanishing quality (QR = 0; e.g. agriculture on rocks), no yield (Y= 0) will be achieved. Second, increasing QR or LA while keeping the other value constant will increase Y. Exactly these two (very weak) properties define the 'qualitative product' in the QDE-formalism.

Unfortunately, the input of labor is not only the condition for production of yield but is at the same time often the cause for the degradation of the natural resource base. Therefore, the influence of LA on the *change* in QR has to be considered. Assuming prevailing agricultural techniques, the resource quality in general decreases faster, if the intensity increases. Conversely, at relatively low intensity even a *recovery* of the natural resource base may occur (e.g., the growth of secondary forests during fallow periods). The intensity which ensures the use of the natural resource without depletion will again depend on several natural and management-dependent factors which are difficult to determine. In accordance with the QDE concept we only assume that such a specific value of LA exists. We call this value *ms* (maximum sustainable intensity). One has to keep in mind that this value depends on the specific division of LA in quality preserving and short-term exploiting measures.

In order to illustrate the functioning of the QDE-formalism let us assume that the simple system described above, i.e., constituted by QR, Y and LA, faces an exogenous determined continuous increase in LA. In this case the formalism deduces three different qualitative time developments ('behaviors').

In Figure 11.3, the possible solutions are shown in detail for QR, Y and LA. Each plot depicts the time behavior of one variable with time on the abscissa and the qualitative value on the ordinate. Thus, the *labor input* in trajectory 1 stays above the value of *ms* for all times. In addition, the arrows indicate the direction of change (see above), i.e., in this example it is also increasing over all times.

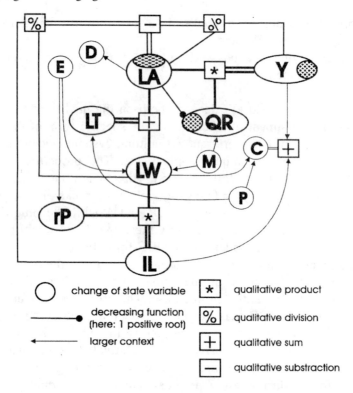

Figure 11.2 Qualitative model for typifying and integration of smallholders' coping strategies into a single set of rules

Variables explicitly included in the qualitative model: LT: total amount of labor; LA: labor investment in agricultural activities; LW: labor investment in wage sector; QR: quality (and quantity) of natural resources; Y: physical yield from agricultural sector; rP: price for agricultural produce relative to present wages; IL: income from wage labor; C: consumption per capita. *Important external variables*: M: socio-economic marginalization of smallholders; P: population; E: national economy; D: Degradation of nature without direct feedback on the resource base.

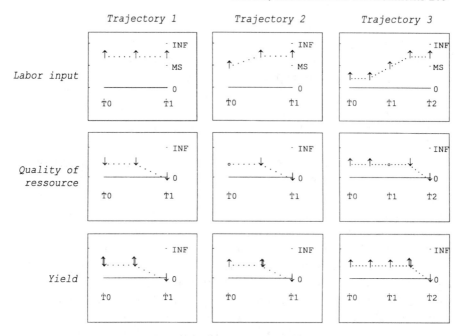

**Figure 11.3 All different solutions of the qualitative model if no wage
labor is present and labor input into agriculture is
continuously increased**

The result mainly depends on the starting value for LA. If one starts
with an input greater than *ms*, QR is reduced throughout at all times
(trajectory 1). Due to the increase of LA and the simultaneous decrease
of QR, the direction of Y is undetermined (double-arrows) until the
resource is exhausted (QR = 0 at T1) which enforces vanishing yield.
Starting with LA = *ms* yields a similar result while in the case of an
initial value less than *ms*, at first a recovery of the resource (increase of
QR) can be observed resulting in an increasing yield over the first time
interval. While LA passes the *ms*-landmark, the recovering trend of QR
is reverted, which leads finally to a similar situation as for trajectory 1.

Allocation of Labor

The next step in the extension of the simple smallholder model is to
introduce the possible division of the total labor potential (LT) into
agricultural (LA) and wage labor (LW) activities which, on the other hand,
means that LW and LA sum up to LT. The resulting total income (C) consists

of the labor income and the income from the agricultural sector (direct consumption, barter, sold at the market). The averaged price relations between the agricultural and the wage labor sector (rP) can be described by the wage per labor unit relative to the mean price of agricultural produce.

In the next step, we include in the model a hypothesized rule of how labor is allocated in the smallholder economy. In an economistic approach, the optimal allocation would be considered assuming the full knowledge of the production system or expressing the uncertainties by probability distributions, and finally optimizing an appropriate goal function, e.g., the integral of the discounted yield over a given time period. It is disputable whether the rules of traditional production systems in their traditional environments approximate such an optimal resource management. Yet, it seems reasonable to assume that under the present unused conditions, smallholders are dealing with a new type of option-box. A relatively widespread criterion for the continuous evaluation of labor allocation decisions is the comparison of the mean outcome per labor input. If it turns out from the comparison that the smallholder is better off in shifting the allocation, it is assumed that he will do so. In the case of our aggregated allocation scheme here, this means to compare the outcome per labor unit of agricultural and the wage labor and reallocate the total available labor in the direction of the more labor efficient activity.

The qualitative model thus developed is a specific part of the overall Sahel Syndrome network of interrelationships where we summarized the complicated dynamics of the natural resource base by only one aggregated trend combining quality and quantity of the natural resource. In Figure 11.2, additional interactions are shown (thin lines) which are relevant for the syndrome but can be interpreted as external in the context of this study. First, some management induced damage of nature (D) which does not directly feed back on the smallholder production system (e.g., loss of biodiversity) has to be considered. Further, we would have to take into account aspects of marginalization (M) leading to restricted access to productive natural resources or wage labor. Next, the influence of population (P) as a source for labor, on the one hand, and divisor for income, on the other hand, should be taken into consideration. Finally, the national and international economic development (E) as determining prices and availability of wage labor has to be reckoned with.

Within the analytic framework of qualitative differential equations we might consider these variables as hidden determinants for the more specific form of those qualitative relations used in the model. For example, all else being equal, population increase will lead to a decrease in consumption. Yet independent of the actual value of P, at present assumed to be constant, any increase of yield Y increases consumption.

Are Important Features of the Case Studies Reproduced?

In Figure 11.4, the structure of all developments which are in accordance with this qualitative model of smallholder labor allocation, income and resource degradation outlined above is displayed. Each marker represents one (numbered) specific state of the system. A state is given by the *qualitative value* and the *direction of change* for each of the variables. The state of the system evolves in time along the branches of the tree. For example, state 1454 has three possible successor states, i.e., 1455, 1446 and 1457. Which of these states in the dynamic development actually follows on state 1454 is not determined by the current formulation of the qualitative net cause-effect in Figure 11.2. Therefore, all possible successor states have to be taken into account. Such ambiguities are the price to pay for the purely qualitative nature of the model.

Specifying the initial condition by requiring that none of the values should be zero or infinity, a total of nine consistent initial states is possible. These states are the ones represented by the first symbols at the left hand side of the tree. The symbols on the right end of each of the nineteen behaviors possible in total indicates whether (a) a jump to another state in the tree eventuates to continue the system development (short diagonal line, the number of the target state is indicated after equal sign), (b) a jump to a previous state on the own trajectory occurs (small circle), or (c) an equilibrium (dot) is achieved.

The first observation is that the dynamics of the system is dominated by cyclical behavior, i.e., almost all branches end with a jump to another state. Only two ends of branches are indicated as equilibria with one of them (no. 1593) being unstable.

The prevailing cyclical behavior is mainly due to the reactive allocation strategy of the smallholder, i.e., when labor productivity of agriculture declines due to decreasing resource quality, the labor input into this sector is reduced, if wage labor is relatively more attractive. The reduction of agricultural activity results after some time in a recovery of the resource. This makes the agricultural activity attractive again which can induce reallocation of labor into the agricultural sector. The frequency of the oscillatory behavior depends on the characteristic time of resource deterioration and recovery which may vary between years and decades. Qualitative modeling cannot resolve the time scale problem as the time points in the solutions of QDEs are just ordered symbols, i.e., time *point* T4 is later than T3 which itself is later than the *interval* (T2,T3). Time points T1, T2, etc. are not divided into equal time slices, but defined by 'events', i.e., characteristic changes in the behavior of at least one system variable. In this sense the event 'resource quality stops decreasing' determines one time point.

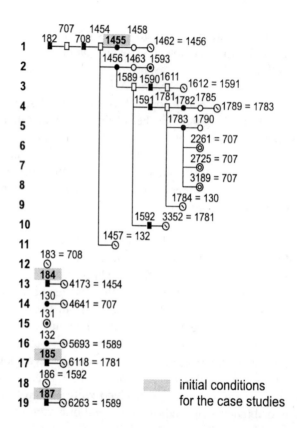

Figure 11.4 Complete behavior tree of the smallholder model

In Figure 11.5, two possible behaviors of the system ('qualitative trajectories') are shown in detail. The state variables displayed are the labor allocated to agriculture (LA), to wage labor (LW), the quality of the natural resource (QR), and the total consumption (CT). Both trajectories start with state 187 (line 19 of Figure 11.4).

In the following sub-section, we discuss how the strategies described in the contributions are reconstructed by the model in form of different branches of the tree.[7] The results as discussed below are summarized in Table 11.5 which also includes an entry describing the future prospects of the regions according to the model output.

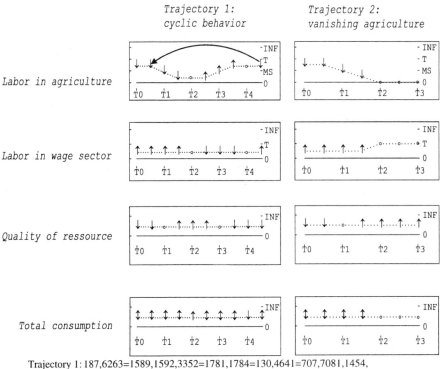

Trajectory 1: 187,6263=1589,1592,3352=1781,1784=130,4641=707,7081,1454,
          1457=132, 5693=1589 à cycle.
Trajectory 2: 187,6263=1589,1592,3352=1781,1783,1790.

Figure 11.5 Two typical behaviors starting from an initial condition compatible with the case study situation in Nepal

Nepal The initial state 187 characterized by declining quality, decreasing labor in agriculture and increasing activities in off-farm labor, can be assigned to the case study on Nepal (chapter 2). The mixed strategy observed among upland farmers there is obviously consistent with the applied qualitative model. It should be noted that this is not as minor as one might suppose since there are only nine consistent initial states and, thus, the fact that one of them coincides with the study of Nepal is remarkable.

Inspection of all the trajectories starting from this initial condition shows that a constant coexistence between wage labor and agricultural labor is *not* possible. Due to the internal dynamics of the agricultural production system, e.g., sustained soil degradation, different forms of oscillations occur. Trajectory 1 in Figure 11.5 shows one such oscillation without intermediate complete allocations of labor into one sector, i.e., it preserves the mixed strategy at all times. It further avoids the temporal, though reversible complete loss of the resource base. This seems to be a positive prospect and should be supported by appropriate policy measures. It means counteracting the total withdrawal from wage labor, keeping wage labor attractive and avoiding incentives to extent agricultural activities. Yet, the development of the total income remains unclear. This, in addition, implies that the time course is very sensitive to the exact form of the interrelations, except that there is no dramatic decline at all.

Table 11.5 Overview of present states in the case studies (numbers of chapters) and of near term prospects (i.e., possible successor states)*

Case study number	Present state in the model				possible successors				
	No.	LA	LW	QR	No.	LA	LW	QR	criticality
(2), (7)	187	>ms,↓	↑	↓	1590	>ms,↓	↑	0	high
					1591	ms, ↓	↑	0	medium
					1592	ms,↓	↑	°	low
(10),(4),(5)	184	>ms,↑	↓	↓	1455/6	(T,°)	0	↓	high
					1457	>ms,°	°	↓	medium
(8; Bela Vista)	1455	(T,°)	(0,°)	↓	1593	>ms,°	0	0	high
					1589	>ms,↓	↑	↓	medium
(8; Dois Irmãos)	185	<ms,↓	↑	↑	1782/3	0,°	T,°	↑	low
					1784	<ms,°	<T,°	↑	low
(3, Ban Kouay)	Traj 1, t=T0	>ms,↑	0	↓	Traj 1, t∈(T0,T1)	>ms,↑	0	0	high
(3, Ban Taohai)	Traj 2, t=T0	ms,↑	0	°	Traj 2, t∈(T0,T1)	>ms,↑	0	↓	medium
	Traj 3, t=T0	<ms,↑	0	↑	Traj 3, t∈(T0,T1)	ms,↑	0	°	low

* The first four cases refer to the behavior tree of the complete allocation model as shown in Figure 11.4. The last two cases, one of which is not unique, are related to the simple model as given in Figure 11.3.

Trajectory 2 in the figure is the only possible time behavior which exhibits a stable total consumption, i.e., this stability is possible only by the total retreat from agriculture. This holds true under the assumption of relatively stable wages, which, of course, has not necessarily to be the case.

Eastern Cape Province (South Africa) Before migrating to Cape Town, the situation of households in rural areas was characterized by decreasing wage labor due to a discriminating Apartheid system (chapter 4).

Accordingly, the activities in subsistence agriculture have been increased, accompanied by a decline in natural resources. This, again, is represented by the initial state 184. Along the trajectory following the states 1454, 1456, 1463 and 1539 the natural resource base is completely diminished at the end. This corresponds to vanishing consumption and, thus, emigration (not modeled here) is the only way out. This actually has happened in South Africa.

Eastern Miombo Highlands Here, it is reported (chapter 5) that in the case of Malawi 'farm operations imply any transition from subsistence production with occasional tobacco market relation and wage labour to petty and extended (tobacco) commodity production'.

This has been translated to the corresponding entry in Table 11.4 and corresponds to state 184. Here, also the 'soil mining' character of tobacco has been used to indicate a decreasing quality of soils - see Table 11.5. The remark given at the end of the contribution, i.e., that if soil mining continues as it is the production potentials of agriculture and the natural environments will be undermined is represented by the trajectory ending in the final state 1593.

Rural Botswana The remarkable governmental drought relief measures in Botswana (chapter 7) can be mapped onto a supported increasing wage labor, decreases of both agricultural activities and resource quality. Again, initial state 187 describes this in terms of the qualitative model. Thus a similar argument holds as given above for Nepal. Yet, the prospects are better here in so far as the governmental measures and the overall better economic situation of Botswana provides a good chance for a promising future.

Dominican Republic The study about the charcoal burners in the Dominican Republic (chapter 10) characterizes their present situation by decreasing wage labor over the last few years. This is reported as a result of a switch from cattle to crop farming which is less labor intensive. Furthermore, the wood-workers have increased their collection activities, here expressed by LA and a documented decline in the natural resource base. This situation is reconstructed by the initial state 184 (line 13 of Figure 11.4) followed by state no.1454 (first line in Fig. 11.4). The GTZ-project *Proyecto Bosque Seco* is reported to largely relief the pressure on natural resources. The project wants to introduce more sustainable methods of forest management, e.g., the use of dead instead of 'green' wood. Within our qualitative model this *exogenous* project can be expressed by increasing the labor threshold for non-sustainable resource use ms, i.e., a constant labor input which is non-sustainable (LA > ms) before the implementation of the new management strategy may turn into a sustainable use (LA<ms). Such an externally induced change in ms transforms to a 'jump' in the behavior tree (Petschel-Held, Block, Cassel-Gintz, Kropp, Lüdeke, Moldenhauer, Reusswig and Schellnhuber, 1999). The case of the *Proyecto Bosque Seco* corresponds to a 'jump' from state 1454 to state 707 which is characterized by LA less than ms, an increasing QR and all other variables equalling those in state 1454.

This 'jump' allows one to evaluate the further development by application of the behavior tree. It shows that the system will inevitably run once more into the undesirable state 1454 because there are no bifurcations in between. So the measure improves the situation but after some time the internal dynamics of the system will reach the old non-sustainable development stage. This corresponds to the situation that due to the positive income effect induced by the project, more and more woodworkers would like to join. In the current model it is assumed that the total available labor is *above* the maximal sustainable yield ms, thus, if finally all woodworkers take part the work will be unsustainable. Therefore, we can say that the measure can be unremitting success only if total labor is less than ms.

Atlantic rainforest, Southeast Brazil   Here, together with the contribution on Laos, two distinct dynamic states are identified according to the two major villages investigated in each study.

In the village of Bela Vista described in the study on Brazil (chapter 8), a constant and high input into agriculture was accompanied by a decreasing resource base and decreasing income. This corresponds to state 1455 which in Figure 11.4 was not explicitly indicated as an initial state. For the second village, Dois Irmãos, decreasing LA, increasing LW and increasing resource quality was found. This represents a positive development which is represented by the initial state 185. By tracing the further development of Bela Vista (1455) the undesirable situation remains until state 1456 is reached. Here a bifurcation occurs which will lead the village either into the disastrous situation of totally exhausting natural resources accompanied by a total decline in income (state 1593) or via 1598, 1592, 3352 to the present state of Dois Irmãos. Conclusively, if at the bifurcation point sufficient wage labor is provided, there is a real chance that Bela Vista can follow the positive path of Dois Irmãos.

Laos In this study (chapter 3), wage labor is generally not a significant alternative to agriculture. Therefore, we have to switch from the labor allocation model which was the basis for the discussions so far, to the results of the simple agricultural model described at the beginning of this section.

Ban Kouay, the first of the villages studied in Sangthong District, shows increasing agricultural activity related *inter alia* to population growth accompanied by decreasing resource quality. This is identified with the initial state (i.e. at time T0) of trajectory 1 - see Figure 11.3. As can be seen, the development of yield (Y) is underdetermined due to the counteracting effects of increasing input and decreasing resource quality. However, the trajectory terminates with totally depleted natural resources and vanishing yield.

In Ban Taohai, on the other hand, increasing agricultural activity is associated with non-decreasing (i.e., constant or increasing) resource quality which can be approximated by the initial state of trajectory 2 or 3, both leading to an increase in yield. But even this positive development in yield is endangered under a steady increase in LA. After LA has passed the critical *ms*-value, the yield increase will return to its former value with a certain time delay, because the instant decrease of QR is masked by the LA increase for some time. Therefore, evaluation of the agricultural system only on the basis of yield development may drive the system in dangerous and almost irreversible states.

Conclusions

The current article has striven to integrate the various contributions contained in this volume. It has been shown that due to the high levels of interlinkages between regions (both due to social and natural processes) it is necessary to relate local developments to Global Change. To this end, it was suggested to perform a pattern analysis of Global Change in order to bridge the gap between specific local aspects and global processes. The integration and comparison of the different contributions has been carried out in the custom of this *pattern identification process*.

It has been shown that all the contributions contain some elements of the Sahel Syndrome representing a major pattern of Global Change. All regions exhibit a serious degree of marginality of agricultural land-use. In many regions marginality is sensitive to climate change.

Using a qualitative modelling approach, it was possible to show that, in spite of dissimilarities in the role of the state, large scale companies or international actors, the coping strategies of smallholders, in principle, obey the same set of basic rules. The centerpiece of this *coping rule* is a rational decision between wage labor and agricultural (farming, herding, collecting) activities. Furthermore, it shows that the diversification of labor allocation between these options is in favor of a sustained livelihood within the region. Otherwise emigration might appear as the only way out of the degradation-impoverishment-spiral.

Coming back to the three basic questions which relate local coping strategies with Global Change (and which have been raised in the introduction), the following conclusions can be drawn.

First, and as it has been supposed in the beginning, the regional settings described in the volume contribute to Global Change through their high levels of soil degradation. Globally, this is important as it represents a significant impact on the global food security system. Locally, there is any reason to assume that in some of the regions smallholders cut the ground under their own feet. Exceptions with good prospects of escaping from increasing soil degradation can be found in Botswana and the Dominican Republic. We can assess the overall criticality by using Table 11.5. We recognize that, independent of the actual path taken, a significant risk exists for the Southwest of the Dominican Republic (chapter 10), the Eastern Cape Province (chapter 4) and the Eastern miombo highlands in Southeast Africa (chapter 5). The situation, however, is not so clear for Laos (chapter 3) and the Atlantic rainforest in Southeast Brazil (chapter 8) since the results differ within the regions studied, i.e., there are good and bad prospects. Similarly, the situation turns out to be ambivalent in the case of Nepal (chapter 2) and Botswana (chapter 7), i.e., development is on the borderline and depends strongly on the near future.

Second, there are several impacts of global processes upon regional or local strategies. In Laos and the Southeast African miombo highlands, the effects of international dam building companies and, respectively, the global tobacco market are of particular relevance.

Table 11.6 Assessment of overall risks by combining the results from climate sensitivity* and local dynamic prospects*

	Positive View	Negative View
Hotspot	Laos	Laos
		Cape Province
		Miombo
		Dominican Republic
Mixed	Laos	Nepal
	Cape Province	Botswana
	Miombo	Atlantic rainforest
	Dominican Republic	
Good	Nepal	Atlantic rainforest
	Botswana	(Southeast Brazil)
	Atlantic rainforest	

* As for climate sensitivity, see Table 11.3. As for local dynamic prospects, see Table 11.5. Due to limited information, no assessment was done for the Tanzanian Maasailand and Huancar District in Argentina. When necessary, the assessment for Laos and Brazil has been split up according to the two villages investigated in each of the case studies. The more critical situation relates to the villages assessed as more critical in Table 11.5.

Third, the natural productivity of all the regions is sensitive towards climatic change. As described above, those regions where marginality is moderate and sensitivity to climate change is medium to high are in a highly critical state. We, thus, can combine this measure with the dynamic prospects (discussed first) to name particular hotspots of future evolution, i.e., the increasing risk due to climate change coincides with an endangerment due to local dynamics. Due to the somewhat ambiguous dynamical prospects, we take two different perspectives, i.e., a positive one where we assume that the dynamical behavior takes the better path, and a negative one assuming the worst possible evolution - see Table 11.5. Regions are classified as 'hotspots' if both contributions, i.e., climate sensitivity *and* the local dynamics, have to be assessed as critical under the respective world-view. In this case, a strong synergetic effect has to be suspected which might dramatically endanger the livelihood of local people. If the situation is not so clear, i.e., with one aspect assessed as critical and the other as rather unproblematic, the overall region is classified as 'mixed'. In such a case, more detailed studies have to follow the assessments as provided in Tables 11.3 and 11.5. Finally some regions were grouped as rather unproblematic since none of the two indicators indicates degrees of criticality.

In total, the analysis has shown that the widespread notion, i.e., that in socio-economic terms the tropics would be worse off if climate changes, does not hold true in the form of a generalized statement. A more elaborate analysis will be necessary to carve out the prospects for different regions. The current analysis is seen to be just a first, highly aggregated step towards attaining this goal. Further work has to be done in order to refine the qualitative modelling of humankind/nature interactions.

Notes

[1] Henceforth, we refer to the different case studies by simply using the corresponding chapter numbers.

[2] Actually, all case studies collected in this volume are related to the question of land-use and land-cover. Some, but not all of them, do further refer to *changes* in land-use as a major factor of human intervention into nature.

[3] Note the medical terminology. Yet, this similarity should not be taken too literally, as, for example, the method for syndrome formulation and validation differs quite a lot between medical science and Earth System Analysis (Schellnhuber and Wenzel, 1998).

[4] Note the similarities of the Sahel Syndrome as compared to different types of the degradation-impoverishment-spiral (Leonhard, 1989; Kates and Haarman, 1992). Nevertheless, the syndrome tries to go beyond the spiral itself and more so models basic mechanisms in more detail.

[5] Formally, sensitivity is defined here as modulus of the gradient of agricultural marginality within a 36-dimensional climate space (monthly mean temperatures, monthly precipitation sum and monthly solar radiation). The resulting values are then classified into three classes, i.e., low (0-200), medium (200-400) and high (larger than 400). The values are taken relative to a monthly change of temperature of $5.13°C$ and of 21.5 mm in precipitation.

[6] If not stated explicitly, the evaluation refers to special case studies as reported in the single contributions of this volume, i.e., on a varying (and partly small) number of villages and households selected and investigated in larger detail. In some cases, more general information on the respective study regions could be relied upon.

[7] The case studies on the Tanzanian Maasailand and Huancar District in Northern Argentina had to be excluded since the model applied is not capable of adequately mapping the conflicts between herders and farmers lying at the heart of the first study, while the second study contains no information on the development of natural resource properties but more so focusses upon ethnological matters.

References

Alcamo, J., Kreileman, E., Krol, M., Leemans, R., Bollen, J., van Minnen, J., Schaefer, M., Toet, S. and de Vries, B. (1998), 'An Instrument for Building Global Change Scenarios', in J. Alcamo, R. Leemans and E. Kreileman (eds), *Global Change Scenarios of the 21st Century*, Pergamon Press, Oxford.

Blaikie, P. and Brookfield, H. (1987), *Land Degradation and Society*, Methuen, London, New York.

Cassel-Gintz, M., Lüdeke, M.K.B., Petschel-Held, G., Reusswig, F., Plöchl, M., Lammel, G. and Schellnhuber, H.J. (1997), 'Fuzzy Logic based Global Assessment of the Marginality of Agricultural Land Use', *Climate Research*, vol. 8, pp. 135-50.

German Scientific Advisory Council on Global Change (1993), *World in Transition: Basic Structure of Human-Nature Interaction, Annual Report 1993*, Economica-Verlag, Bonn.

German Scientific Advisory Council on Global Change (1994), *World in Transition: The Threat to Soils, Annual Report 1994*, Economica-Verlag, Bonn.

German Scientific Advisory Council on Global Change (1997), *World in Transition: The Research Challenge, Annual Report 1996*, Springer-Verlag, Berlin.

German Scientific Advisory Council on Global Change (1999 a), *World in Transition: Ways towards Sustainable Management of Freshwater Resources, Annual Report 1997*, Springer-Verlag, Berlin.

German Scientific Advisory Council on Global Change (1999 b), *World in Transition: Strategies for Managing Global Environmental Risks, Annual Report 1998*, Springer-Verlag, Berlin.

German Scientific Advisory Council on Global Change (1999 c), *World in Transition: Annual Report 1999*, Springer-Verlag, Berlin.

Haken, H. (ed) (1983), *Synergetics*, Springer-Verlag, Berlin.

Intergovernmental Panel on Climate Change (1996), *The Science of Climate Change: Second Assessment Report*, Cambridge University Press, Cambridge.

International Commission on Large Dams (1984), *World Register on Large Dams: Full Edition*, ICOLD, Paris.

International Commission on Large Dams (1988), *World Register on Large Dams: Update*, ICOLD, Paris.

Kasperson, J.X., Kasperson, R.E. and Turner, B.L.II (eds) (1995), *Regions at Risk: Comparisons of Threatened Environments*, United Nations University Press, Tokyo, New York, Paris.

Kasperson, R.E., Kasperson, J.X., Turner, B.L.II, Dow, K. and Meyer, W.B. (1995), 'Critical Environmental Regions: Concepts, Distinctions, and Issues', in Kasperson, J.X., Kasperson, R.E. and Turner, B.L.II (eds), *Regions at Risk: Comparisons of Threatened Environments*, United Nations University Press, Tokyo, New York, Paris, pp. 1-41.

Kates, R.W., Turner, B.L.II and Clark, W.C. (1990), 'The Great Transformation', in B.L. Turner II, W.C. Clark, R.W. Kates, J.F. Richards, J.T. Mathews and W.B. Meyer (eds), *The Earth as Transformed by Human Action*, Cambridge University Press, Cambridge, pp. 1-18.

Kates, R.W. and Haarman, V. (1992), 'Where the Poor Live: Are the Assumptions Correct?', *Environment*, vol. 34, pp.4-11, 25-8.

Kuipers, B. (1994), *Qualitative Reasoning: Modeling and Simulation with Incomplete Knowledge*, MIT Press, Cambridge, MA.

Leonhard, H.J. (1989), *Environment and the Poor: Development Strategies for a Common Agenda*, Transaction Books, New Brunswick, Oxford.

Little, P.D. and Horowitz, M.M. (eds) (1987), *Lands at Risk in the Third World: Local-Level Perspectives*, Westview Press, London.

Lüdeke, M.K.B., Moldenhauer, O. and Petschel-Held, G. (1999), 'Rural Poverty driven Soil Degradation under Climate Change: the Sensitivity of the Disposition towards the SAHEL SYNDROME with respect to Climate', *Environmental Modelling and Assessment,* (forthcoming).

McCully, P. (1996), *Silenced Rivers: The Ecology and Politics of Large Dams*, Zed Books, London.

Nicolis, G. and Prigogine, I. (1977), *Self-Organization in Non-Equilibrium Systems*, Wiley, New York.

Petschel-Held, G., Block, A., Cassel-Gintz, M., Kropp, J., Lüdeke, M.K.B., Moldenhauer, O., Reusswig, F. and Schellnhuber, H.J. (1999), 'Syndromes of Global Change: A Qualitative Modeling Approach to Assist Global Environmental Management', *Environmental Modelling and Assessment,* (forthcoming).

Schellnhuber, H.J. and Wenzel, V. (1998), *Earth System Analysis: Integrating Science for Sustainability*, Springer-Verlag, Berlin, Heidelberg, New York.

Schellnhuber, H.J., Block, A., Cassel-Gintz, M., Kropp, J., Lammel, G., Lass, W., Lienenkamp, R., Loose, C., Lüdeke, M.K.B., Moldenhauer, O., Petschel-Held, G., Plöchl, M. and Reusswig, F. (1997), 'Syndromes of Global Change', *GAIA: Ecological Perspectives in Science, Humanities and Economics*, vol. 6 (1), pp. 19-34.

Stern, P.C., Young, O.R. and Druckman, D. (1992), *Global Environmental Change: Understanding the Human Dimensions*, National Academy Press, Washington, DC.

Strengers, B.J., den Elzen, M.G.J. and Kösters, H.W. (1997), 'The Land and Food Submodel: TERRA', in J. Rotmans and B. de Vries (eds), *Perspectives on Global Change: The TARGETS Approach,* Cambridge University Press, Cambridge, pp. 135-58.

Turner, B.L.II, Clark, W.C., Kates, R.W., Richards, J.F., Mathews, J.T., and Meyer, W.B. (eds) (1990), *The Earth as Transformed by Human Action,* Cambridge University Press, Cambridge.

Turner, B.L.II, Skole, D., Sanderson, S., Fischer, G., Fresco, L. and Leemanns, R. (1995), *Land-Use and Land-Cover Change: Science and Research Plan,* IGBP-Report No. 35/HDP-Report No. 7, Stockholm, Geneva.

Zobler, L. (1986), *A World Soil File for Global Climate Modeling*, NASA Technical Memorandum No. 87802, Goddard Institute for Space Studies, New York.

Zuidema, G., van den Born, G.J., Kreileman, E. and Alcamo, J. (1994), 'Simulation of Global Land Cover Changes as affected by Economic Factors and Climate', in J. Alcamo (ed), *IMAGE 2.0: Integrated Modelling of Global Climate Change,* Kluwer Academic Publisher, Dordrecht, Boston, London, pp. 163-98.

Index of Authors

References from Notes indicated by 'n' after page preference.

Index of Subject Matters

References from Notes indicated by 'n' after page preference.